Making Universal Service Policy

Enhancing the Process Through Multidisciplinary Evaluation

TELECOMMUNICATIONS

A Series of Volumes Edited
by Christopher H. Sterling

Borden/Harvey • The Electronic Grapevine: Rumor, Reputation, and Reporting in the New On-Line Environment

Bracken/Sterling • Telecommunications Research Resources: An Annotated Guide

Brock • Toward a Competitive Telecommunications Industry: Selected Papers From the 1994 Telecommunications Policy Research Conference

Brock/Rosston • The Internet and Telecommunications Policy: Selected Papers From the 1995 Telecommunications Policy Research Conference

Cherry/Wildman/Hammond • Making Universal Service Policy: Enhancing the Process Through Multidisciplinary Evaluation

Lehr • Quality and Reliability of Telecommunications Infrastructure

Mody/Bauer/Straubhaar • Telecommunications Politics: Ownership and Control of the Information Highway in Developing Countries

Noll • Highway of Dreams: A Critical View Along the Information Superhighway

Regli • Wireless: Strategically Liberalizing the Telecommunications Market

Rosston/Waterman • Interconnection and the Internet: Selected Papers From the 1996 Telecommunications Policy Research Conference

Teske • American Regulatory Federalism and Telecommunications Infrastructure

Waterman/MacKie-Mason • Telephony, the Internet, and the Media: Selected Papers From the 1997 Telecommunications Policy Research Conference

Williams/Pavlik • The People's Right to Know: Media, Democracy, and the Information Highway

Making Universal Service Policy

Enhancing the Process Through Multidisciplinary Evaluation

Barbara A. Cherry
Michigan State University

Steven S. Wildman
Michigan State University

Allen S. Hammond, IV
Santa Clara University

LAWRENCE ERLBAUM ASSOCIATES, PUBLISHERS
1999 Mahwah, New Jersey London

Lawrence Erlbaum Associates, Inc., Publishers
10 Industrial Avenue
Mahwah, NJ 07430

Cover design by Jennifer A. Sterling

Library of Congress Cataloging-in-Publication Data

Making Universal Service Policy: Enhancing the Process Through Multidisciplinary Evaluation/ [edited by] Barbara A. Cherry, Steven S. Wildman, and Allen S. Hammond, IV.
p. cm. (Telecommunications)
Includes bibliographical references and index
ISBN 0-8058-2456-1 (alk. paper). —ISBN 0-8058-2457-X (alk. paper)
1. Telecommunication—Law and legislation—United States. 2. Telecommunication policy—United States. I. Cherry, Barbara A. II. Wildman, Steven S., III Hammond, Allen IV 1950– . Series: Telecommunications (Mahwah, N.J.)
KF2765.U55 1999
343.7309'94—dc21

98—7805
CIP

Books published by Lawrence Erlbaum Associates are printed on acid-free paper, and their bindings are chosen for strength and durability.

Printed in the United States of America
10 9 8 7 6 5 4 3 2 1

To Hayden for his patience and support
—B.A.C.

To Paul and Arletha for Susan,
who makes it so much easier
—S.S.W.

To George Gerbner, Henry Geller, and Michael Botein,
who believed in me and gave me a chance
—A.S.H. IV

Contents

List of Contributors xi

Preface xiii

I: INTRODUCTION 1

1 Conceptualizing Universal Service: Definitions, Context, Social Process, and Politics 3
Barbara A. Cherry and Steven S. Wildman

II: FRAMEWORKS FOR ANALYZING UNIVERSAL SERVICE 13

2 Universal Service: Migration of Metaphors 15
Harmeet Sawhney and Krishna Jayakar

3 Unilateral and Bilateral Rules: A Framework for Increasing Competition While Meeting Universal Service Goals in Telecommunications 39
Barbara A. Cherry and Steven S. Wildman

4 Questions for Outlining a Universal Service Policy 59
Marlin Blizinski

III: SOCIETAL ROLE AND IMPLICATIONS OF UNIVERSAL SERVICE 67

5 Rethinking Universal Service: What's on the Menu? 69
Marlin Blizinski and Jorge Reina Schement

6. The Social Architecture of Community Computing **85**
Allen W. Batteau

7 Universal Access to Infrastructure and Information **99**
Allen S. Hammond, IV

IV: PAYING FOR UNIVERSAL SERVICE **109**

8 Overview of Universal Service **111**
Carol Weinhaus, Robert K. Lock, Harry Albright, Mark Jamison, Fred Hedemark, Dan Harns, and Sandra Makeeff

9 Recovering Access Costs: The Debate **135**
David Gabel

10 Universal Service: A Stakeholder Response **159**
James C. Smith

V: EMBARKING ON A NEW UNIVERSAL SERVICE POLICY: THE ROLE OF THE FEDERAL GOVERNMENT **165**

11 Review of Federal Universal Service Policy in the United States **167**
Barbara A. Cherry and Steven S. Wildman

12 Some Legal Puzzles in the 1996 Statutory Provisions for Universal Telecommunications Services **179**
Warren G. Lavey

13 Universal Service and the National Information Infrastructure (NII): Making the Grade on the Information Superhighway **189**
James McConnaughey

VI: THE ROLE OF THE STATES **213**

14 The New State Role in Ensuring Universal Telecommunications Services **215**
Thomas W. Bonnett

15 Breaking the Bottleneck and Sharing the Wealth: A Perspective on Universal Service Policy in an Era of Local Competition **237**
Robert K. Lock, Jr.

Author Index **251**

Subject Index **257**

List of Contributors

Harry Albright, Manager, Federal Regulatory Planning & Policy, Ameritech

Alan Batteau, Professor, Department of Anthropology, Wayne State University

Marlin Blizinsky, Manager, King County Office of Cable Communications, Seattle, Washington

Thomas W. Bonnett, Director of Economic Development & Environment, Council of Governors' Policy Advisors

Barbara A. Cherry, Associate Professor of Telecommunication and Associate Director of the James H. and Mary B.Quello Center for Telecommunication Management and Law, Michigan State University; formerly Adjunct Professor, Department of Communication Studies, Northwestern University, and Director of Public Policy Studies, Ameritech

David Gable, Professor, Department of Economics, Queens College

Allen Hammond, IV, Professor of Law, University of Santa Clara Law School

Dan Harris, District Manager, Separations & Capital Recovery, Bell Atlantic

Frank Hedemark, Director, Government Affairs, AT&T

Krishna Jayakar, Doctoral Student, Department of Telecommunications, Indiana University

Warren G. Lavey, Partner, Skadden, Arps, Slate, Meagher & Flom, Chicago, Illinois

Robert K. Lock, Jr., Senior Vice-President, Competitive Strategies, Chicago, Illinois; formerly Executive Assistant to Commissioner Dirkson in Illinois Commerce Commission

Sandra Makeef, Staff, Iowa State Utilities Board

James McConnaughey, Senior Economist, National Telecommunications and Information Administration

Harmeet Sawhney, Associate Professor, Department of Telecommunications, University of Indiana

Jorge Schement, Associate Dean for Graduate Studies and Research, College of Communications, Pennsylvania State University

James C. Smith, Regulatory Vice President, Ameritech Ohio

Carol Weinhaus, Director, Telecommunications Industries Analysis Project, Public Utilities Research Center, University of Florida

Steven S. Wildman, James H. Quello Chair of Telecommunication Studies and Director of the James H. and Mary B. Quello Center for Telecommunication Management and Law, Michigan State University; formerly Associate Professor of Communication Studies and Director, Program in Telecommunications Science, Management & Policy, Northwestern University

Preface

The public policy of universal service in the field of telecommunications has existed for decades, but its meaning and implementation have always generated debate. However, the debate has been particularly contentious and the issues more complex in recent years, given advancements in technology and regulatory reform toward more competitive provision of telecommunications services. In the United States, universal service policy was expressly codified by Congress for the first time in the Telecommunications Act of 1996; and the Federal Communications Commission (FCC) and state public service commissions are struggling to implement the numerous Congressional mandates.

This volume is the outgrowth of shared interests between the editors and the contributing authors to provide a multidisciplinary perspective for evaluating universal service policy and recommending policy changes to accommodate a more competitive telecommunications environment. Too often recommendations for universal service policy are based on a single discipline, such as economics or law, or simply political expediency. The importance of telecommunications technology and availability today requires more insightful analysis and a multidisciplinary approach.

In the quest to formulate a multidisciplinary approach, the editors first agreed to organize a conference, soliciting the input and participation of individuals from disparate academic disciplines; local, state, and federal government bodies; and industry members. This conference, bearing the name "Universal Service in Context: A Multidisciplinary Perspective," was held December 6, 1995, at the New York Law School's Communications Media Center.

Most of the chapters in this volume are based on papers that were presented at the conference. However, during the editing process of this book, significant changes were made in the policy arena—not the least of which was the passage of the Telecommunications Act of 1996 and ensuing recommendations by the Federal–State Joint Board on Universal Service in November 1996. In determining when to submit the manuscript, the editors struggled to determine when the manuscript would be timely. We determined ultimately to await the FCC's implementation order on universal service, which was issued in May 1997. As a result of both the Joint Board recommendations and the subsequent FCC order, we decided that the volume required modifications. Some chapters were revised to reflect the recent developments, and new authors were sought out to contribute additional chapters in order to provide a more balanced collection of views. The resulting volume reflects recent significant developments in U.S. universal service policy, the implementation of which continues to unfold.

ACKNOWLEDGMENTS

We have many people to thank for their support over the past 2 years as this book took shape. The financial support of Ameritech, The Benton Foundation, MCI Communications, the Communications Media Center at New York Law School, and the Program in Telecommunications, Science, Management and Policy at Northwestern University was essential for the 1995 conference from which this book developed. As for the administrative support of the conference, Roberta Tasley of the Communications Media Center was indispensable.

We owe a great debt to the authors of this book, who contributed substantial time and talent in drafting their respective chapters. We are grateful to Theomary Karamanis for computer support in preparation of the manuscript and her assistance in bringing the project to closure. We also thank Linda Bathgate of Lawrence Erlbaum Associates for her patience and encouragement of the editors' efforts to bring this book to fruition.

Our special thanks to our supportive spouses and children, without whose understanding the completion of this book would not have been possible.

—Barbara A. Cherry
—Steven S. Wildman
—Allen S. Hammond, IV

I

INTRODUCTION

1

Conceptualizing Universal Service: Definitions, Context, Social Process, and Politics

Barbara A. Cherry
Northwestern University

Steven S. Wildman
Northwestern University

Official universal service policies, as reflected in laws, regulatory statutes, and budgetary appropriations are, most immediately, products of government. In this sense, it is appropriate to say they are, at any moment in time, products of conscious attempts to create mechanisms to promote the attainment of those policy goals that held sway when they were created. From this perspective, universal service policies are developed. However, the goals of policy emerge from the interplay of social, economic, and technological forces that influence and are influenced by government. These forces, and the directions in which they push policy, change over time and vary with circumstances that differ among nations. From this perspective it is appropriate to say that universal service policies evolve as manifestations of underlying forces more basic than government. In the case of universal service in the United States, this evolution is reflected in repeated attempts to come up with new definitions for universal service to respond to ever-advancing technological capabilities and a sense of social entitlement that advances with them.

The debate over how universal service is defined has focused on the terms and conditions under which narrowly specified bundles of services are made available to end users. Because government reacts to the political realities of the moment and then responds with a lag, we find that the political consensus and technological justifications for a new definition of universal service are already evaporating as it is being implemented. This, in turn sets the stage for the political battle over the next definition that is to guide universal service policy. The current battle over the scope of telecommunications services to be provided to schools, libraries, and health care providers under Section 254 (h) of the 1996 Telecommunications Act is an example of this policy merry-go-round. In the May 8, 1997 Report and Order in Docket No. 96-45, the Federal Communications Commission (FCC) mandated provision of Internet access and internal connections for these institutions—although most observers feel that the FCC's inclusion of this requirement as a component of universal service was a response to political pressure, and that such inclusion exceeds the FCC's authority under the 1996 Act.

Continual strife among stakeholders is inevitable as long as we persist in defining universal service policy goals in terms of prices for access to specific bundles of services, all of which are destined to become obsolete as technology continues to advance. A central theme among the chapters in this volume is that the process by which universal service policies are developed would be more effective if universal service policies were conceived of as time-dependent outcomes of a political process driven and constrained by social, technological, and economic forces. In the long run, information technology's contribution to social welfare would be greater if we worked to devise a process for guiding the evolution of universal service that was more enduring than any particular set of policies that might be implemented along the way. This requires a broader perspective on what universal service is and should mean than has been evident up to this point. This volume contributes toward the development of that broader perspective.

Although the approach to universal policy advanced here is general, the individual chapters take as a backdrop the development and evolution of universal service policies in the United States. Universal service has been an increasingly contentious topic of debate in telecommunications policymaking since the 1984 breakup of the Bell System ushered in an era of competition. That the political debate over universal service became increasingly heated as telecommunications competition (or at least its prospect) intensified is a direct reflection of the financial stakes involved. Whatever its flaws, the predivestiture Bell System represented a universal service policy solution with fairly broad political support. Although business-to-residence and urban-to-rural support flows may have been criticized as economically inefficient, this arrangement was supported by a stable coalition of political interests. Because the system was one of regulated monopoly, the system was not subject to the pressures of contending industry groups attempting to use the political and regulatory processes to promote their own competitive advantage. AT&T earned a stable and healthy profit and it was generally believed that residences and rural customers received service at lower rates than would have been offered in an unregulated market.

This all began to change with the introduction of competition, first in long-distance service and more recently in local services. Profits, and often economic viability, were directly impacted by the regulatory decisions on who paid how much to whom that were unavoidable components of universal service policies. Competition created a new set of stakeholders and increased what was at risk in regulatory decisions for old stakeholders. Not surprisingly, all stakeholders, both new and old, increased their spending on lawyers, consultants, and lobbyists to try to mold universal service policies to their advantage. Increased politicization of universal service was thus an inevitable consequence of competition in telecommunications. Divergent economic interests have also made fashioning a political consensus around a new set of policies increasingly difficult.

The attempt to articulate new universal service goals and develop new support mechanisms is further complicated by the fact that rapidly advancing technologies have been changing dramatically the range of current and potentially available telecommunications services and lowering the costs at which they can be supplied. Concomitantly, the debate over universal service has enlarged to include the question of what services should be universally available—that is, just what is meant by the word "service" in universal service is now seen as up for grabs. As new communication capabilities arise and then decline in cost, political pressures develop to include them in the package of socially guaranteed services. Rapid technological change also characterized the predivestiture era, but this was largely reflected in declining costs of transmission and switching, while the set of services available to end customers remained fairly stable. For this reason, it was easy to accept, virtually without debate, that universal service meant access to dial tone voice service. The debate was over the terms of access to a very simple and basic service that, for most customers, changed only gradually as service quality improved over time.[1]

The rapidity with which new services have been added to the list deemed worthy of universal service support in the last several years makes unambiguously clear how quickly the political system can respond to technology-driven opportunities by enlarging the definition of universal service. The FCC's recent decision to include Internet access and internal connections as part of universal service support for schools, libraries, and health care providers is an example. Nevertheless, the official policy debate over the definition of universal service is still being conducted as if the objective were to decide on entitlements to a static bundle of well-defined information services. Although at odds with the process by which universal service policies are actually set, this approach may have been serviceable when, for the most part, only a single basic service was at issue and it changed very slowly. This approach is out of step with the rapid change in

[1]The slow changeover from operator assistance to automated switching for local calls is an example of the gradual qualitative improvements in basic voice service that characterized much of the predivestiture era.

the nature and scope of telecommunications services in the current industry, and it constitutes a serious impediment to effective policy design.

Universal service is a sociopolitical construct, and we need to find a way to design universal policy around a conceptualization of universal service as the product of social and political forces that respond to a specific, but always temporary, set of economic and technological possibilities. As these possibilities change, the definition of universal service inevitably changes with them. Building on a recognition of this basic political truth, a more effective process for policy design might seek agreement on how entitlements are to be changed as circumstances change, rather than what they should be at any given moment.

To reconceptualize universal service in this way—as process-determined—is also to see it as context-determined. In particular, the meaning and import of universal service are determined by the many varied contexts in which people use information technologies and services. So universal service policy evolves as a politically mediated response to changes that occur in these contexts. If policy is viewed as the means by which society attempts to accomplish goals that reflect and transcend the wants and aspirations of its individual members, then effective universal service policies cannot be developed without a deep appreciation for the contextual nature of technology use. Effective policy design also requires that fundamental economic and political constraints not be ignored.

This volume promotes the broader appreciation of the multifaceted nature of the contexts that must be understood to develop effective future universal service policies. The chapter authors represent a broad cross section of disciplinary training, professional positions, and relationships to the telecommunications industry. Academic disciplines represented include law, economics, anthropology, communication, and business. Authors include academics; attorneys; consultants; employees of local, state, and federal government agencies; and representatives of regulated telecommunications firms. The volume is organized as follows:

Part II presents three general analytical frameworks for analyzing universal service issues and developing universal service policies. In chapter 2, Sawhney and Jayakar offer a general systems perspective from which universal service is seen as the outcome of the long term social and political processes by which societies respond to the economic possibilities created by new technologies. They begin by modeling the expansion of a generic technological system (from its inception to maturity) as occurring in three dimensions—geographic, demographic, and layered (elaboration of system complexity). This is followed by three detailed case studies of different systems—the telegraph, education, and suffrage—that developed toward universal availability over time in the United States. They show that a varied mix of political, economic, and social forces may drive a system toward universality and that the possible paths leading toward this end are many.

Lessons drawn for universal service policy in telecommunications are: (a) An official definition is not necessary to the development of universal service; (b) Examples of successfully implemented programs, whether in different localities for telecommunications services or for other (nontelecommunications) services, are likely to exert greater influence on the nature and scope of universal service than are formal definitions; and (c) universal service is a bundle of policies that arose from different sources over time and were forged into a more integrated formal policy with local variants.

Chapter 3 turns from the long-term historical, social, and political processes examined by Sawhney and Jayakar to a study of the legal and institutional conditions that must be satisfied if we are to maintain and meet traditional universal service goals while managing a relatively smooth transition from the traditional system of franchised monopoly carriers to one that relies more heavily on the discipline of competitive markets. Cherry and Wildman argue that the universal service requirements imposed on regulated carriers in the past were sustainable because the system of franchise monopoly ensured that, barring incompetence or malfeasance, a carrier's revenues would be sufficient to cover its costs in aggregate—although the activities supporting universal service were not remunerative. Now policymakers are trying to meet traditional universal service goals without having thought through the changes in policy necessary to make these goals attainable in competitive telecommunications markets. Cherry and Wildman show that the carrier obligations associated with many of these goals create vulnerabilities for competitive firms that would make them unsustainable if they were simply imposed as unilateral requirements. In these cases, special financial assurances must be provided by government if the desired services are to be provided by competitive firms. They introduce the analytical framework of unilateral and bilateral rules to help policymakers determine when such assurance is required and identify the least intrusive forms such assurance might take. They then argue that the transition to more competitive telecommunications markets would be facilitated, and in the long term more reliance on competitive processes would be possible, if carriers' constitutional protections against loss of economic value due to government actions were interpreted more broadly.

Blizinski shifts the focus in chapter 4 from the broad concerns about legal and regulatory rules examined in chapter 3 to specific questions that must be asked and resolved through the political process in setting the goals of universal service policy. Ultimately, decisions on the nature and extent of universal policies will reflect considerations of costs and benefits. In assessing the benefits and costs of universal service policies, Blizinski cautions against a too narrowly drawn economic conception of benefits—pointing out that, to the extent that communication technologies facilitate social interaction and political participation, they generate spillover benefits to society that may substantially exceed the sum of direct personal benefits experienced by individual users. If this more expansive view of the benefits side of universal service is accepted, it also justifies a more inclusive perspective on who should contribute to universal service support and the reasons for doing so. Blizinski concludes with a discussion of the roles different levels of government should play in formulating and implementing universal service policies. To the extent that national uniformity in services is viewed as important, then a stronger federal role is warranted. However, if universal service is viewed as an instrument for addressing needs and opportunities that vary locally, then state and local governments should be given more discretion in implementing universal policies.

The three chapters in Part III examine social and political contexts in which universal service policies impact communities and the lives of individuals. In chapter 5, Blizinski and Schement consider universal service "a tool for enabling citizens to participate in the fundamental activities of society, including its economic, political, and social life." They present demographic statistics showing that the poor, the young, minorities, and families headed by females with children living at or below the poverty level are disproportionately represented among the 6% of Americans not connected to the public-switched telephone network. A close look at the reasons individuals and families choose not to take telephone service shows factors other than ability to pay are often determinative in this decision. On the basis of their demographic analysis, they argue that demographic changes and an expanding menu of media choices warrant a rethinking of universal service goals. They recommend expanding universal service to include cable television, telephony, and the Internet, and that individuals be allowed to choose from a menu of media options.

From the demographics of media consumption, chapter 6 turns to how individuals, groups, and communities use communication technology to serve their particular needs. Observing that we "lack an understanding of the demand characteristics and potential uses for advanced information technology and resources in the inner city," Batteau reports the findings of two studies of community computing based on efforts to introduce computers and local networking in low income inner-city neighborhoods. The most salient information needs identified in the communities studied related to finding and getting jobs, education, and housing. He discusses the relationship between local network architecture and the social structure and needs of communities. Using job-related information needs as an example, he describes a set of services that could be provided with relatively outdated computer and networking technology and more sophisticated services that might be supplied as technology advances.

Political participation is more central to the analysis of chapter 7, where Hammond observes that, as we change the telecommunications infrastructure and the services and information it makes available, we simultaneously change what it means and takes to be an informed and responsible citizen, a well-educated individual, and a skilled, in-demand employee. Of particular concern is the possibility that inner-city urban and rural communities are not participating fully in the advancing infrastructure, creating a class of "information have nots." Hammond sees a complex of interrelated factors contributing toward this outcome, including a workforce prepared inadequately for the high technology jobs of an increasingly information-oriented economy, schools that are equipped inadequately to address this need, and an information infrastructure that is older and less capable of providing advanced services than ones in more affluent areas. He expresses concern that such needs will not be addressed adequately by an information policy largely focused on competition and argues that, in making policy decisions affecting what advanced services will be available to all citizens, we are really making long-term decisions about what we will be as a nation.

The cost of universal service is the subject of Part IV, with three chapters addressing various issues relating to policy costs. Chapter 8, by Weinhaus, Lock, Albright, Hedemark, Harris, and Makeeff, provides a statistical and descriptive overview of the current system of support, the primary components of which are local access payments by interexchange carriers, rural–urban price averaging, and more explicit and targeted support programs for rural telephone companies. They discuss the cost of this system of support—including the various approaches that have been advanced for estimating such costs, proposals and their cost implications for changing this system, and the sociopolitical barriers to making serious change. Weinhaus et al. argue that the combination of existing government telecommunications policies favoring competition and changing technologies are making the old system of support unsustainable. They provide estimates of the impact of rate deaveraging on rural penetration and examine benchmark cost models as a device for estimating the magnitude of subsidies required to support rural service. They also consider mechanisms for phasing in a new system of subsidies and the cost implications of expanding the definition of universal service to include broadband services.

Gabel offers a very different perspective on subsidy in chapter 9. His focus is narrower than that of Weinhaus et al. and restricted to the question of how much of the cost of access to the local exchange should be recovered in the price of local service and how much should be recovered through access charges paid by long-distance carriers. Because the loop is utilized for both local and long-distance calling and loop costs do not vary with calling volume, its costs are fixed and common to both services. A critical question in the debate over how much contribution to loop costs should be built into long-distance prices through access charges paid by interexchange carriers is how large this contribution would be in a competitive unregulated telecommunications market. Also debated is the extent to which fixed costs should be recovered in fixed versus usage-based charges to customers. Gabel examines AT&T's pricing policies during the historical era when it competed with independent telephone companies for customers in local markets to show that prices charged different classes of customers (business vs. residential, urban vs. rural) implicitly involved considerable sharing of common network costs, including internal transfers among customers and services that in the current debate would be termed subsidies. Furthermore, internal AT&T documents show that this price structure was consciously designed to produce this outcome. Gabel argues that competitive markets do not necessarily recover fixed costs in fixed fees by showing that in markets for other services, such as credit cards, usage charges often contribute to the recovery of fixed costs.

An executive with Ameritech, Smith offers the perspective of a regulated local exchange carrier on the recovery of universal service costs in chapter 10. Although acknowledging the economic benefit of promoting more extensive subscription to the public-switched network, Smith points out that much of the support for universal service goals has been provided in the form of subsidies that are implicit in the structure of regulated prices and requirements imposed on incumbent carriers that are sustainable only in a closed system of regulated monopoly. He suggests that, although growing competition in telecommunications markets is making traditional mechanisms for supporting universal service increasingly untenable, the universal service goals supported by those mechanisms are still achievable if we move to a system of explicit subsidies supported by a set of competitively neutral mechanisms for funding them. Smith recommends a three-step approach that would clearly identify the goals motivating past universal service policies and determine which of those still require support. He suggests that such an approach to revamping universal service policy would result in a scaling back of government programs and discusses the lessons to be learned from various reregulatory initiatives affecting universal service in the Ameritech region states.

Parts V and VI focus on recent and ongoing developments in universal policy at the federal and state levels of government, respectively. In chapter 11, Cherry and Wildman provide a broad overview of federal universal service policy. The chapter provides a concise summary of the relevant sections of the 1996 Telecommunications Act, lists the specific universal service policy goals and policies enunciated in the Act, and reviews the subsequent actions of the FCC and the Federal-State Joint Board to implement sections of the Act relating to universal service. Because there are frequent references to the 1996 Act throughout this volume, readers not intimately familiar with it may find it useful to refer to this chapter from time to time as they read chapters by other authors.

With a piece of legislation as complex and comprehensive in its intended reach as the 1996 Telecommunications Act, there are bound to be ambiguities that will have to be resolved by regulators and the courts in the course of implementation. In chapter 12, Lavey identifies three puzzles in the Act's statutory provisions that he says remain to be worked out by the FCC, state regulators, and the courts. One is that, although the Act requires that the aggregate of federal and state support be sufficient to meet the goals of universal service policy, it provides no mechanisms for coordinating contributions that are authorized separately at the two levels of government. Second, no rules are provided for determining whether federal support mechanisms are sufficient to meet the requirements of universal service, given that the costs of universal service may vary widely in response to economic conditions and regulatory practices that differ substantially from state to state. Third is the requirement that the costs of policies be predictable. Because the outliers that make prediction difficult comprise only a small fraction of local exchange carriers, Lavey recommends that the FCC continue to deal with local exchanges with unusually high costs as exceptions, as it has in the past, while restricting the amount of support to be provided through exception treatment. Sufficiency would thus be defined in terms of providing support sufficient for "the vast majority of areas not receiving exception treatment."

In chapter 13, McConnaughey describes the role the federal government has played in promoting universal service during the Clinton administration, in the process providing a virtual catalog of recent federal studies and policy initiatives related to the promotion of access to the information infrastructure. He discusses the results of universal service and related policies in the past and shows how federal policies are helping to expand the concept of universal service "from dial tone for households to basic information technology access for all Americans." McConnaughey provides demographic summary statistics on the penetration and consumption of a wide range of information products and services that complement those provided by Blizinski and Schement by expanding the range of technologies and services considered. He notes that as we focus increasingly on access to an information infrastructure rather than simple dial tone service, information policy, including universal service, must deal with issues relating to access to public institutions such as schools and libraries and the access these institutions can provide to other sources of information.

Because government responsibility for supporting and meeting universal service goals is split between the states and the federal government, it is not possible to understand universal service policy in the United States without a clear understanding of what is going on in the states. Bonnett provides a review of state universal service policies and initiatives in chapter 14 that parallels the review of federal policies by Cherry and Wildman in chapter 11. Although the pace of deregulation has varied considerably among the states, a number of states made considerable progress toward revising rates and regulatory procedures to bring them into line with the requirements of competitive telecommunications markets prior to the Telecommunications Act of 1996. Now these efforts must be coordinated with the requirements of the Act. Bonnett chronicles these efforts and discusses the challenges the states face in bringing their policies into harmony with federal initiatives under the Act.

Harmonization of telecommunications policy across regulatory jurisdictions is also a focal issue in chapter 15 by Lock, which concludes the volume. Lock observes that the transition to competition has made it necessary for policymakers to reevaluate their prior assumptions about universal service—including the terms of the regulatory contract under which universal service policy was administered under franchise monopoly. The fact that federal agencies and 50 state governments are pursuing their own initiatives makes harmonization difficult. He offers four principles to guide this process of reevaluation: (a) Rates should be allowed to more closely reflect costs; (b) Regulators must take the time required to develop the records of evidence needed to have some confidence in the effectiveness of new policies; (c) Policies must be tailored to the characteristics of individual jurisdictions; and (d) Regulatory policies must incorporate monitoring mechanisms that facilitate flexible responses to varying circumstances. The U.S. experience with state reform of telecommunications regulation has been one in which a few states have been early leaders and innovators, with policy initiatives in follower states incorporating in part the innovations and experiences of the first movers. Lock is concerned that the innovations of states that lead the charge into reregulation are being adopted too uncritically by those states that started the reform process later—with the result that genuine differences among states are not being adequately reflected in their plans for reregulation. These recommendations and insights are supported by a review of regulatory reform in Illinois.

II

FRAMEWORKS FOR ANALYZING UNIVERSAL SERVICE

2

Universal Service: Migration Of Metaphors

Harmeet Sawhney
Indiana University
Krishna Jayakar
Indiana University

Metaphors are the building blocks of our psychological reality. From the tenuous foothold of our limited knowledge, we shape the confusing and encircling morass of everyday actuality into neat metaphoric bricks of understanding and build them up into the edifices of a worldview. We use metaphors to bridge the chasm between the frontiers of knowledge and the unknowns of new phenomena.

Nowhere is this more apparent than in our efforts to make sense of the dawning information age. As we take tentative steps toward formulating policies for new information and communication technologies, we are confronted with critical decisions. With the unprecedented rate of technological and social change, and with little data to fall back on, our efforts are little more than the gropings of individuals in an unfamiliar, dark room. All we have are metaphors, which mold our thoughts, guide our instincts, and shape our actions.

The importance of metaphors in shaping our policies for advanced telecommunications makes it necessary for us to reexamine the appropriateness of the metaphors we use. According to Fiumara (1995), *metaphoricity* is a basic mode of functioning whereby we project patterns from one domain of experience in order to structure another domain of a different kind. Metaphoric thought involves the discovery of recurring patterns of relationships between elements in two different regions of experience, denoted as the principal subject and the subsidiary subject (Brown, 1976; Gentner & Jeziorski, 1993; Nuessel, 1988). "A metaphor selects, emphasizes, suppresses, and organizes features of [its] principal subject by implying statements about it that normally apply to the subsidiary subject" (Nuessel, 1988, p. 14). Thus, the metaphor allows us to make sense of a new and unexplored phenomenon by selecting and organizing its observed features into patterns familiar to us from our experience with similar phenomena (Black, 1962). The specific linguistic term employed in metaphor is only shorthand for the recurring pattern of relationships observed between the two subject areas. Therefore, metaphors are best evaluated not on the basis of the explicit terminology used in the metaphor, but at the deeper level of the implicit experiential gestalts invoked by the metaphor (Johnson, 1981).

In the case of universal service, our thinking has not kept pace with technological change. We continue to invoke an experiential gestalt drawn from a bygone regulated monopoly era while technological change has altered the rules of the game. Our efforts to develop appropriate universal service policies have attained limited success because we have sought to make modifications within the existing paradigm. We need a new starting point for our thinking. One of the best ways to do this is to adopt a new metaphor that provides a more relevant framework for our thoughts.

This chapter explores other domains of experience—telegraph, universal suffrage, and universal education—to identify metaphors that would provide the appropriate experiential gestalt for the challenge we currently face with universal service. We first conceptualize the three modes of system expansion—territorial, demographic, and layered—that push the system toward ubiquity and increasing complexity. We then use examples from telegraph, universal suffrage, and universal education to illustrate these three modes. Finally, we locate the specific case of universal service in advanced telecommunications within the larger framework and draw on the relevant experiential gestalts to suggest new policy directions.

SYSTEM EXPANSION: A CONCEPTUAL FRAMEWORK

This section discusses three modes of system expansion—territorial, demographic, and layered. These modes are discussed in conceptual terms without reference to specific examples, which are provided in the next section.

Territorial Expansion

Territorial expansion is the extension or replication of systems across geographical space over a period of time. Extension implies connectivity, which in turn implicates proximity or contiguity as important factors in territorial expansion. Extension is generally outward from a center to the periphery. However, territorial expansion can also take place by replication in geographically discontinuous areas (see Fig. 2.1). Proximity is thus not defined exclusively in geographical terms. Multiple centers may exist in territorial expansion. Under such circumstances, territorial expansion also passes through a stage of system integration. Replication may be partial or

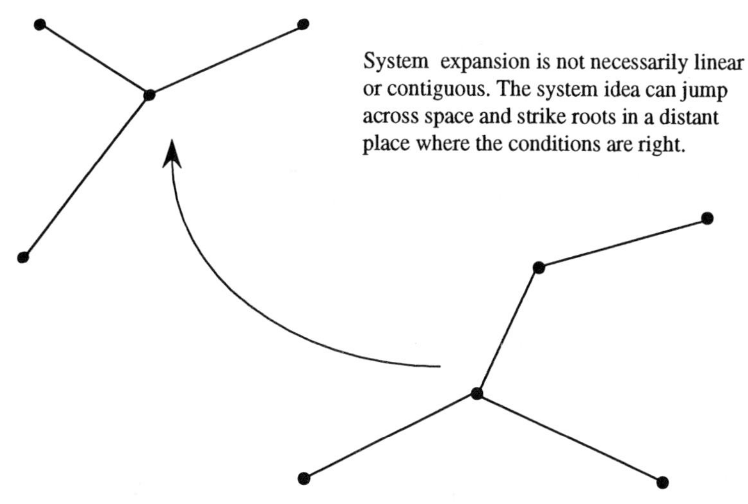

FIG. 2.1. Territorial expansion.

total, depending on the degree to which the new system is imitative of those already in place.

Space in territorial expansion is directional because a place is defined not by its absolute location but by its relative position in the systemized space, which in turn influences the source of innovations it adopts. For example, the former West and East Germanies, although geographically contiguous, looked west and east respectively and belonged to different policy spaces. Also, space is hierarchized as center-periphery or as node-trunk-hub-spoke-terminal in networked systems. These positions are not fixed; a terminal position in a networked system may be transformed into an important hub if it becomes the point of interconnection to a neighboring networked system.

Demographic Expansion

Demographic expansion takes place within the population at each location via a process involving in groups and out groups.[1] The process expands the definition of in groups participating in a system to include progressively larger sections of the population over a period of time (see Fig. 2.2). Early participants or adopters form the core group, to which new groups are pro-

---

[1]The terms in groups and out groups have a long history in identity politics, cultural studies, and other related areas. Not all the connotations of these terms are invoked by their usage here. We have used these loaded terms partly for want of better terminology, and partly to highlight the differences in the interests and perspectives of those who have access to a service or resource and those who do not.

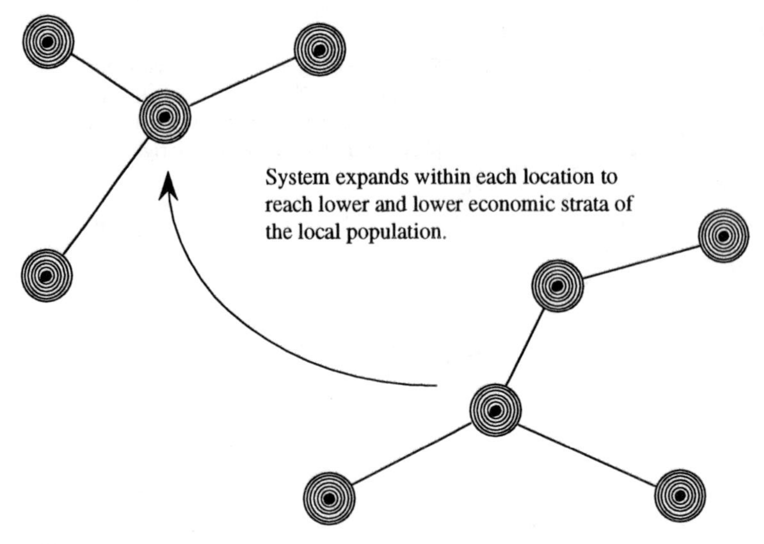

FIG. 2.2. Demographic expansion.

gressively added either by expanding the scope of the definition or by including new categories in the definition.

Demographic expansion can be either intentional or incidental. In intentional demographic expansion, membership in the in group is based on formal criteria, and the definition is expanded by an authority empowered to do so. In incidental demographic expansion, there are no formal requirements for membership, and the in group is expanded gradually through participation. Incidental demographic expansion is the closest to the conventional description of diffusion in the literature.

Layered Expansion

Layered expansion is the asynchronous addition of a hierarchy of complementary innovations to existing systems by which they are continually modified and made more complex. Innovations are asynchronous because their time of initiation is different, and they have different penetration rates. No two adoption units are confronted with adoption decisions for the different innovations in the same order. Innovations are hierarchized because they can be arranged on a scale of increasing complexity and the adoption of some innovations has to precede that of others (see Fig. 2.3). Innovations are complementary because they belong to the same policy area, influence one another, and their progressive adoption can be understood in terms of movement toward a general policy goal or social objective. In that sense, innovations in layered expansion are modular. Here, the adoption decision is more complex than in territorial or demographic expansion as

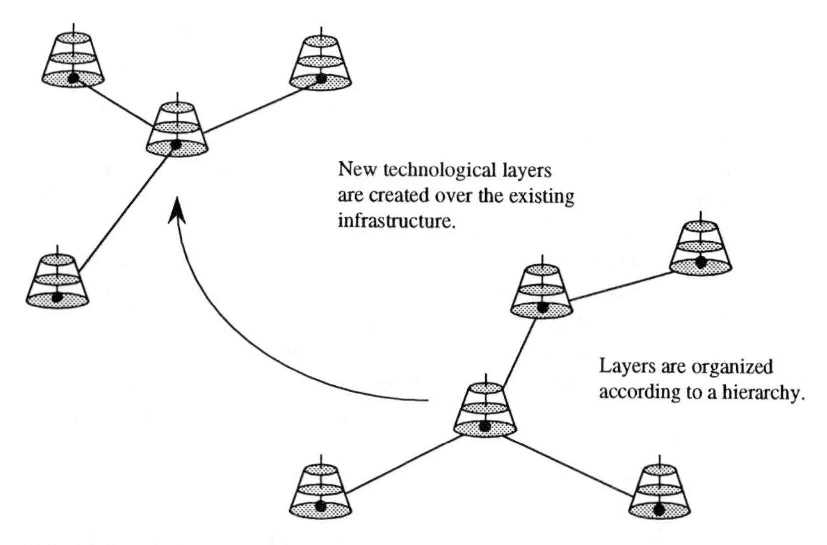

FIG. 2.3. Layered expansion.

adopters have to make decisions at a set of points, not just about each innovation, but about the combination of innovations to be included in the policy bundle.

The hierarchy of innovations in layered expansion is divided between the essential and the optional. Innovations on the lower rungs of the hierarchy are designated as essential, whereas other innovations on the higher levels are optional. The process of layered expansion involves the progressive upward movement of the level of innovations considered essential. Once a lower level is standardized and integrated into the system, there is competition among the alternatives in the next higher level for inclusion. The system expands by choosing one competing alternative and eliminating or restricting the others.

It is important to note that the three modes of system expansion do not necessarily take place in any one particular order. Under some circumstances, it is possible for a system to pass through more than one of the expansion processes just described, either simultaneously or in varying sequential combinations.

EXPANSION PROCESSES: THREE ILLUSTRATIONS

This section provides three examples that illustrate the modes of system expansion just discussed. They are the growth of the telegraph network, universal suffrage, and universal education.

Telegraph

The Morse telegraph network started commercial operations in January 1846 on a line from New York to Philadelphia, and gradually expanded its coverage. It evolved into a confederation of local telegraph companies that were interconnected with one another (Brock, 1981). The typical arrangement was that the local franchisee gave Morse 50% of the stock for the right to use the patented technology. The network essentially grew in a hierarchized manner as it moved down the urban hierarchy from big cities to smaller and smaller towns until it reached the outer fringes, which were too sparsely populated to make the investment profitable. Gradually, a fairly extensive network developed in the populated areas of the eastern seaboard. Then a quirk of fate unleashed forces that propelled a dramatic expansion of the network in a brief 5-year period from 1847 to 1852 (Thompson, 1972).

It all began with a rather innocuous franchise agreement between Morse and Henry O'Reilly for the construction and operation of a telegraph network covering the territory west of Philadelphia lying between the Ohio River, the Great Lakes, and the Mississippi. Because the Morse interests did not expect O'Reilly's line to generate much revenue, they agreed to a reduced patent fee—only 25% of stock. O'Reilly's line turned out to be a major success. The Morse interests regretted the fact that they had relinquished control over a vast territory for half the standard patent fee and found a loophole via a technicality to nullify the contract. O'Reilly, once a partner, became a fierce competitor and the ensuing rivalry set into motion the "mad era of methodless enthusiasm" (Thompson, 1972, p. 97). The end result was a reckless race to occupy virgin territory. Network investment decisions were not made on the basis of potential profitability but the burning desire to be the first to establish a telegraph network in new territories. The network exploded and soon it was covering the continental expanse of the United States.

Telegraph's growth was essentially one of territorial expansion. It did not involve either demographic or layered expansion as the telegraph never became individualized like suffrage, nor household-based like the telephone. Here, the primary mode of network expansion was extension rather than replication.

Universal Suffrage

The territorial expansion of universal suffrage was particularly evident in the United States because of its federal structure, with the states taking the important decisions in enfranchising residents. As the population expanded westward, the states created in newly settled territories often replicated the democratic institutions and practices of the more established eastern states. This replication was typically partial as the new states modified the institutional forms and practices they borrowed to fit their peculiar environmental factors, thus serving as crucibles for the emergence of new variations, which in turn, influenced the eastern states. There was considerable transfer of ideas along both directions as the United States moved towards universal suffrage.

The expansion of universal suffrage also had a demographic dimension. Universal suffrage was neither granted to Americans by benevolent founding fathers, nor was it bestowed in any single act of great magnanimity. At first, the voting rights were restricted to the propertied elite, the in group. Adult franchise was extended to the rest of the population, the out group, in a very piecemeal manner as a result of a protracted struggle between competing political interests within the in group.

A democracy in many ways is a game of numbers in which the side with the largest number of votes emerges as the winner. In a situation of limited democracy where the suffrage is restricted, the numbers game can be influenced by expanding the population base of eligible voters. This possibility of manipulation has frequently led to fraudulent practices. When voting privileges were restricted to freeholders, "the practice of creating freeholds at the time of a crucial election was fairly widespread in the U.S. colonies. When an office seeker and his friends thought an additional number of votes was necessary to win an election, it was not unusual for them to create small freeholds for the express purpose of manufacturing votes" (Williamson, 1960, p. 50). The other strategy was to ease the restrictions on suffrage so that a more partisanly disposed population could be included in the voter pool. In Pennsylvania, the Quaker party used the naturalization of aliens as the means of enlarging its voter base (Williamson, 1960). The midwestern states extended suffrage to foreigners for economic reasons to "attract immigrants from other states to her unplowed fields" (Porter, 1969, p. 18). The unrolling of universal suffrage was a series of compromises and political compacts within the in group that had very little to do with the espoused ideals of freedom and the equality of man. These negotiated interim compromises offered temporary advantages to different players and hence provided motivations for extending suffrage. Each step forward opened new doors and thereby had a snowballing effect. This political maneuvering among the subgroups within the ever-increasing in group eventually led to suffrage being extended to each and every adult citizen of the United States. The entire experience has been described aptly as the "strange phenomenon of suffrage being carried forward on a tide of fallacies and specious doctrine" (Porter, 1969, p. 65).[2]

---

[2]This discussion on the demographic expansion of universal suffrage has been extracted with minor modifications from Sawhney (1994).

The movement toward universal suffrage stopped at demographic expansion because it had a definite closure—one person, one vote. Universal suffrage by its very definition could not accommodate qualitative differences in the voting privileges it granted citizens. Therefore, layered expansion was not an issue in the case of universal suffrage.

Universal Education

The system of universal education as we know it is a bundle of different policies including compulsory attendance laws, sources for school finance, provision of schools and colleges in territorial jurisdictions, mandatory standards for teachers and syllabi, and enforcement mechanisms for all of these elements. The evolution of this complex system involved all three modes of expansion—territorial, demographic, and layered—that often occurred simultaneously in an intertwined manner. This nonsequential system development is evident in the following discussion, which traces the growth of universal education since the early 17th century.

The first immigrant groups settled the continent's eastern seaboard in relatively isolated and independent colonies with distinct national and religious identities. The educational initiatives of the colonies were more or less independent of each other. After independence, federalism ensured that each state within the union enjoyed considerable freedom to choose the educational system most suited to itself. However, their independent identities did not prevent colonies (and later states) from borrowing ideas from each other. Successful innovations were copied from other territorial jurisdictions, especially those regarded as exemplars. Exemplars were usually the states considered more progressive, as well as the ones with which the adopting state had the most in common. Thus, exemplars were often states from the same region as the adopting state. For instance, there were major differences between the educational initiatives of New England and the southern colonies. In New England, concern with the religious instruction of the young and the republican ideal of the equality of all citizens prompted community elders to create the early statutes making education compulsory for all children. The colonies of Massachusetts and New Haven were the exemplars for colonial New England (Kotin & Aikman, 1980). Their laws were copied by all the New England colonies except Rhode Island between 1642 and 1671, making it a mandatory requirement for all children to be taught by their parents, or in the case of apprentices by masters.

Educational requirements were gradually expanded from a prescribed minimum to incorporate new and higher level requirements. For example, the first Massachusetts law of 1642 requiring all children to be taught reading was amended in 1648 to say that pupils should be able to read perfectly (Kotin & Aikman, 1980). In 1660, the New Haven colony added a writing requirement.[3] As each colony adopted educational statutes, it had the option to devise new requirements, or to borrow from more than one source. The Plymouth colony's education law of 1671 illustrates this. It borrowed the educational requirements from Massachusetts, and enforcement provisions from the New Haven law.[4] Significantly, it did not have any requirement that children should be taught a trade, which both the Massachusetts and New Haven laws had, demonstrating that adoption was not wholesale but quite selective.

At the same time, the southern colonies had Virginia for their exemplar. Notably, no southern colony adopted a law requiring all children to be given education, as was the case in New England. According to Urban and Wagoner (1996), the Virginia model used "every man according to his own ability" (p. 23) as its motto. Because well-to-do families would naturally provide their wards with the best education they could afford, the state needed to concern itself only with those incapable of acquiring an education on their own, namely "orphans, poor children, illegitimate children, and mulattos born of white mothers" (Kotin & Aikman, 1980, p. 19). In 1705, Virginia introduced a law stating that all apprentices should be taught a trade and to read and write. These Virginia laws were copied in the other southern colonies.

Two competing models were thus available for other colonies or states. Ultimately, it was the Virginia model of apprenticeship education that prevailed. Even in New England, universal compulsory education ceased to be a concern after the late-1600s, to resurface only in the 1800s.[5] In the intervening period, education statutes in America referred not to universal education, but to the responsibilities of masters toward their apprentices.

---

[3]As Urban and Wagoner (1996) pointed out, the addition of a compulsory writing requirement was radical for the times. Compared to reading, for which primers and other books were cheap and readily available, writing material was expensive and hard to obtain. It was regarded as a specialized skill enabling its practitioners to enter lucrative professions in law, business, and public affairs. Writing was also taught largely by male teachers, whereas reading was the domain of female instructors.

[4]The enforcement provisions in the New Haven law may be regarded as the first recognition of the rights of the state in the education of the young. The law allowed colonial, rather than local authorities to supervise enforcement of the law; and the negligence of parents or masters to provide a suitable education to their wards was punishable by fines (Kotin & Aikman, 1980).

[5]See Kotin and Aikman (1980) and Hillesheim and Merrill (1971) for some of the reasons why the New England model was gradually eclipsed. First, continued immigration of an ethnically and religiously diverse population diluted the homogeneity of early settlements and made consensus on educational content difficult. Secondly, with the increasing secularization of education, the church, which had taken a leadership role in colonial education, gradually abdicated control to local authorities. Thirdly, as populations moved westward, settlements became too dispersed to support a school system and made compulsory education statutes impossible to implement.

The 18th and 19th centuries were marked by a large number of educational initiatives that flourished for some time, and then died out or evolved into new forms. As Good (1956) pointed out, "one may discern the faint outlines of several systems" (p. 37) in the educational diversity of this period. These systems of education offered alternatives to students with different professional aspirations and economic means. For the common masses there were dame schools, parochial schools, old field schools, and district schools, which prepared children to become tradesmen or clerks.[6] In the cities, there were venture schools and private schools through which pupils could acquire a good education that prepared them for the intermediate professions. In these schools, practical skills like accounting and bookkeeping were imparted with relatively less emphasis on scholarship. Finally, there were the preparatory schools and Latin grammar schools, which prepared future scholars, doctors, clergymen, and lawyers for college. Most of the children in these schools belonged to privileged families.

It should be noted here that the emergence of educational institutions did not follow any sort of graded structure. For example, the first colleges were instituted almost contemporaneously with the first elementary schools.[7] As systems evolved, existing institutions were integrated into a graded structure, so that they formed an educational track or hierarchy. Existing forms were sometimes modified and combined to fill an intermediate position in the evolving hierarchy. An example of this is the combination of the college preparatory grammar school and the terminal venture school into the academy, a secondary level institution. These hybrid institutions combined the desirable features of both predecessor institutions. For example, the academy offered a variety of programs, including both college preparatory and terminal, and was private like the venture school but had the grammar school's rigorous academic standards (Hillesheim & Merrill, 1971).

Another trend that marked the 18th and the first half of the 19th century was the increasing role of the state in education. This was not without opposition from advocates of parental rights or the rights of local communities. But, by the early decades of the 20th century, the rights of the state to set reasonable educational standards had acquired considerable public and judi-

---

[6]As the name indicates, *dame schools* were run by women in their homes where very young children were taught their letters. Old field schools, parochial schools, and district schools were all elementary schools under different forms of management. *Old field schools*, run on the basis of a contract between teacher and parents, got their name from being located on waste or exhausted land belonging to the town. Parochial schools and district schools were run by the church and an elected board respectively (Good, 1956, see pp. 37–38).

[7]Nine colleges were founded in America during the colonial period, with the first, Harvard, as early as 1636 (Hillesheim & Merrill, 1971).

cial sanction.[8] The standard-setting role of the state paved the way for the evolution of statewide and nationwide systems of education. In the absence of this standard-setting role, nascent educational systems would probably have undergone continuous mutation without convergence.

In the early 19th century, a constellation of factors resurrected the old Puritan ideal of education for all in an era marked by industrialization, rural-to-urban migration, and massive immigration. Many believed that the consequent social dislocation would lead to class conflict and discord. Educational reformers like Horace Mann advocated that the school system be used as a means of assimilating diverse ethnic and linguistic groups into a common U.S. identity. As the country industrialized rapidly, there was an increasing demand for better educated, skilled workers. There was also a need to inculcate in prospective factory workers qualities like punctuality, teamwork, and obedience to authority. The school system was an effective means of achieving these purposes.

Also important was the sociological transformation of the U.S. workplace. In the colonial period and for many decades after that, able-bodied workers were difficult to find, and all available hands had to be employed in field or workshop. Within this context, child labor was condoned and encouraged in the name of the Puritan work ethic. Continuing immigration increased the labor supply and thereby reduced the need for child labor. The labor interests realized that the continuing employment of children in the workplace was depressing wages for all. It was therefore necessary to remove a part of the child workforce from the labor pool. Chief Justice Brenner of the U.S. Supreme Court was alluding to this when he referred to "the economic function (of the educational system in) keeping children of certain ages off the labor market and in school." (*Wisconsin v. Yoder*, 1972).

Under these influences, the *common school*—free, tax-supported, nondenominational, and universal—began to generate more and more popular enthusiasm. Universal education gained acceptance as both a desirable ideal and a realizable goal. To effect the transition to a universal and compulsory system of public education, reforms were required on a number of fronts: compulsory attendance laws to compel the presence of all school-age children in educational establishments, child labor laws to pro-

[8]As the United States Supreme Court stated in *Pierce v. Society of Sisters*, " [n]o question is raised concerning the power of the state reasonably to regulate all schools, to inspect, supervise and examine them, their teachers and pupils; to require that all children of proper age attend some school, that teachers shall be of good moral character and patriotic disposition, that certain studies plainly essential to good citizenship must be taught, and that nothing to be taught which is manifestly inimical to the public welfare" (quoted in Katz, 1976, p. 26). It should, however, be noted that the same judgment negated the rights of states to compel the attendance of all children in public schools, provided their parents imparted some form of education to them. *Pierce v. Society of Sisters* examined the constitutionality of a 1922 Oregon statute that compelled public school education for all children 8- to 12-years-old.

hibit children in the workplace, and school finance. Initiatives in all three areas were taken by different state jurisdictions and the innovations diffused rapidly over the union once their utility was established. Rhode Island was the first to initiate a child labor law in 1840 (Alexander & Jordan, 1973). By 1853, six states had child labor laws, and by the end of the century the number had risen to 28 (Kotin & Aikman, 1980). Compulsory attendance laws were first introduced in Massachusetts in 1852, requiring all children aged 8 to 12 to be sent to school for a minimum of 12 weeks per year, with at least 6 weeks running consecutively (Kotin & Aikman 1980). The District of Columbia followed suit in 1864 and Vermont in 1867. By 1900, 38 states had compulsory education statutes, with the southern states enacting their own laws by 1918.

Innovations in school finance were much more controversial because they directly affected the tax liabilities of all citizens. Earlier, schools had been financed by private endowments, land grants, parental contributions, or permanent school funds, created mostly from the sale of land, the income from which was earmarked for schools (Edwards & Richey, 1963). Once the school systems expanded, an institutional arrangement that could manage funds, oversee their disbursement, and regulate their utilization became necessary. In 1812, the state of New York established an office of the superintendent of common schools, adopted a district system, and permitted a district tax for school finance. This system was rapidly emulated by the other states. By 1860, 28 of the 34 states had chief state school officers, and the district system had become firmly entrenched (Edwards & Richey, 1963). The creation of an educational bureaucracy was an impelling force toward the institutionalization of universal education (Katz, 1976). Thus, during the first few decades of the 20th century:

> compulsory schooling was transformed from a relatively simple statute requiring a fixed period of school attendance into a complex network of interrelated rules. This network of rules involved not only requiring school attendance but also the hiring of truant officers, delegating jurisdictive power, and dealing with a host of child labor regulations. (Katz, 1976, p. 21)

Eventually, universal education inched up to the high school level in the early 20th century, and it seemed to have reached a natural plateau. The conventional wisdom was that the next level, college, because of its very nature, could not be prescribed for everyone on a uniform basis. It would have stayed like this for a long time had it not been for World War II. The war effort required a mass mobilization, and the army could not get enough college-trained persons for its officer corps. It therefore used its Army General Classification Tests to identify men who had potential, gave them advanced training, and commissioned them as officers (Bowles, 1966). The success of this program suggested that twice as many persons had the aptitude to attend college as were actually enrolled in the prewar period. This realization created much disquiet in the highest policy circles because it suggested that the nation was not fully utilizing its human potential. Herein lay the seeds for the subsequent initiatives for universal higher education (Bowles, 1966; Munday & Rever, 1971; Willingham, 1970).

Soon after the war, President Truman established the Commission on Higher Education to study the civilian implications of this war experience. The Commission recommended that tuition-free education should be made available to everybody for the first 2 years of study either in traditional colleges or community colleges. It felt that "only in this way can we be certain of developing for the common good all the potential leadership our society produces, no matter in what social or economic stratum it appears" (President's Commission on Higher Education, 1947, p. 38). The postwar success of the G.I. Bill further confirmed the belief that economic barriers were preventing a large number of otherwise-capable persons from fully developing their potential (Bowles, 1966; Munday & Rever, 1971; Willingham, 1970). Further, studies suggested that the G.I. Bill eventually cost the taxpayers nothing because the increased taxes paid by the beneficiaries more than compensated for the original investment (Educational Policies Commission, 1964).

The movement toward universal higher education was given a further boost by a critical view of higher education that started striking roots in the 1960s. It saw the institutions of higher education as reproducing and perpetuating inequality (Arboleda, 1981; Birenbaum, 1971; Lavin, Alba, & Silberstein, 1981; Rossman, H.S. Astin, A.W. Astin, & el-Khawas, 1975). They were viewed as sorting and screening mechanisms that distributed power and privilege by creating and credentialing an elite (Schrag, 1971; U.S. Office of Education, 1971; Willingham, 1970). According to this view, the purpose of higher education should not be to pick winners but to maximize the educational potential of each and every student at whatever level he or she enters the higher education system (Karabel, 1972; Rossman et al., 1975).

The war experience, reports of the President's Commission on Higher Education and other bodies, the advent of the critical view of higher education, and other elements of the sociocultural milieu of the 1960s, including the Civil Rights movement, converged to create pressure for universal higher education. The biggest impediment for any movement forward was a lack of conceptual understanding of what exactly universal higher education entailed. It was yet another ideal that was conceptually powerful but difficult to operationalize. The past experience with secondary education offered few precedents because higher education could not be prescribed on an uniform basis to everyone. Should higher education be made available to everyone or should it be restricted to those who have the required talent? Should it have an uniform curriculum or should it offer great diversity? These questions were not a problem earlier because in the past, progress was linear and incremental as each grade was an additional layer on a curriculum already in place. Universal higher education, on the other hand, represented a major discontinuity. The questions it raised were profound and the answers elusive.

Yet, the lack of definitive answers to these fundamental questions did not stop the evolutionary process. The process continued forward as exemplars emerged and defined what was possible in practical terms. The course of universal higher education was greatly influenced by two very different models represented by the California Higher Education System and the City University of New York. California adopted a stratified strategy of differential access with three levels of colleges—University of California system, California State University system, and a community college system. The upper 1/8 of high school graduates were placed in the University of California campuses, the upper 1/3 in the California State University campuses, and the community colleges were open to everybody with a high school diploma. All students were assured admission to the system but not to a specific college. The system, however, allowed for upward mobility as a student could work his or her way up the hierarchy (Jaffe & Adams, 1972). The City University of New York adopted a very different open admissions model that assured admission to all the high school graduates from New York who ranked in the top half of their graduating class. The resulting increase in student body was accommodated by hiring new faculty and increasing class sizes. Furthermore, a concerted effort was made to offer remedial programs to take care of the needs of the weaker students (Lavin et al., 1981). The other states developed their own arrangements that were hybrids of elements drawn from these two models. The fascinating thing about this whole experience is that all these major policy initiatives were taken within a context of great ambiguity.

The development of universal education followed a more complex evolutionary path than that of either the telegraph or universal suffrage. It required a number of separate but complementary initiatives to be put together over a period of time. Initiatives arose in different jurisdictions at different points in time, and diffused over the federal system at different rates. Educational requirements began with elementary skills and were gradually upgraded to include higher levels. There was competition between alternative forms as new layers were added to the educational system. This layered process makes the development of the universal education system an illustrative example of layered expansion.

THE MIGRATION OF METAPHORS

The influence of metaphors is very evident when they are used explicitly to frame an issue. Their influence is more subtle and powerful when they are so deeply embedded in a discourse that even the participants are not aware of it. For example, Lakoff and Johnson (1980) pointed out how common discourse on "argument," laden with words like "attack," "defense," and "demolish," is structured by the conceptual metaphor "argument is war."[9] We comment, analyze, and actually engage in an argument within the conceptual structure of the metaphor without even being aware of it.

This section performs a similar analysis of the implicit conceptual metaphors that informed the development of the telephone network in the different stages of its growth and thereafter suggest a new metaphor for the future. We do not claim that these metaphors were explicitly used in the development of universal service policies. However, we suggest that, like the "argument is war" metaphor, these metaphors formed the conceptual structure within which different initiatives were undertaken.

As the telephone network evolved, it went through territorial, demographic, and layered modes of expansion. Naturally, the metaphoric terms in which we thought about the network also changed to reflect the dominant processes in the different stages of its expansion. The telegraph metaphor, which was adequate to understand territorial expansion in the early phase of the telephone network, was replaced by a derivative of the universal suffrage metaphor—the universal service principle—when demographic expansion became the dominant mode of network expansion. As layered expansion becomes the dominant process in the evolution of the telephone network, the metaphors we employ will have to be changed again. Figure 2.4 illustrates this migration of metaphors and shows that the universal education metaphor encompasses the layered expansion processes currently active in network expansion.

The first stage of telephone's development was guided by the telegraph metaphor. This was quite natural because the telephone was an unexpected technological mutation that emerged out of efforts to improve the telegraph. Even the patent filed by Bell for his new invention was titled "Improvements in Telegraphy." It did not even mention the word "telephone" (Brooks, 1976). The influence of the telegraph metaphor pervaded all aspects of the telephone business. "Even the language of the telephone revealed its ancestry; telephone calls were for many years labeled as messages and measured in message units" (Fischer, 1992, p. 81). In fact

---

[9]The metaphor "argument is war" can be seen operating in our everyday language as seen in the following statements:

> Your claims are *indefensible.*
> He *attacked every weak point* in my argument.
> His criticism were *right on target.*
> I *demolished* his argument.
> I've never *won* an argument with him.
> You disagree? Okay, *shoot!*
> If you use that *strategy, he'll wipe you out.*
> He *shot down* all of my arguments. (Lakoff & Johnson, 1980, p. 4)

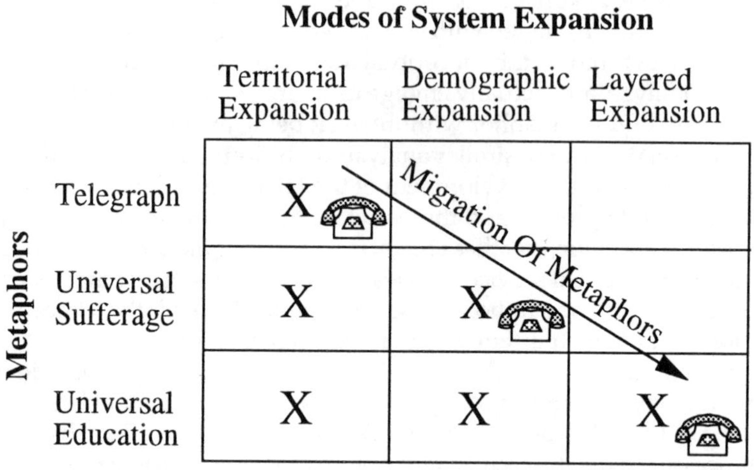

Modes of System Expansion

Metaphors	Territorial Expansion	Demographic Expansion	Layered Expansion
Telegraph	X		
Universal Sufferage	X	X	
Universal Education	X	X	X

Migration Of Metaphors

FIG. 2.4. Migration of metaphors.

many of the key inventions, like telephone exchanges, which facilitated the development of early telephony, grew out of ideas first developed for telegraph (Fagen, 1975; Garnet, 1985; Mueller 1989).

Although the telegraph network expanded mainly by extension, the telephone network grew by both extension and replication. The growth of the Bell System represented the former and the independents the latter. In the beginning, the Bell companies focused almost entirely on the urban markets. As in the case of the Morse telegraph network, the Bell System grew in a hierarchized manner down the urban hierarchy. It refused to serve the rural areas in spite of vociferous demands from the rural population. As soon as the Bell patents expired, independent telephone companies mushroomed all across the rural landscape. They were often crude homegrown telephone networks put together by farmers that "replicated" the Bell networks. The surprising success of the independents forced the Bell companies to enter the rural areas. The introduction of competition radically altered the pace of network development (Mueller, 1993). There was an almost nine-fold increase in the per capita density of telephones in the period 1893–1902 (Fischer, 1987). In this phase of the network's development, the foundations were laid for universal service—a geographically ubiquitous telephone network that interconnected all communities.

The next step in the development of universal service was the extension of telephone service to all citizens. The mode of network development at this stage was very different from the previous one as the extension of the network to "everyone" followed a very different logic from the earlier extension to "everywhere." The extension of the telephone network to everyone required considerable investment in the local loop that connected individual subscribers to the telephone network. This created many problems because the local loop is a dedicated and not a shared facility, and there is a distinct cost for connecting each individual subscriber to the network. This issue did not arise in the case of telegraph because the network was not extended to each individual subscriber. The extension to everyone raised questions that the telegraph metaphor could no longer address. Should society subsidize the local loop for those who could not pay the full cost? The metaphor that helped address this question was that of universal suffrage.[10]

Initially, the telephone service was limited to the rich and the business community—the in group. It was almost a luxury. Slowly, as prices fell, the subscribership continued to expand beyond the small group of wealthy elites to those on the lower rungs of the economic ladder as more people could afford to pay for telephone service. The in group was positively predisposed toward universal service because the expansion of the network increased the economic value of their telephone service as they could communicate with a larger universe of subscribers.[11] The only hitch was that this expansion necessitated a subsidy from the in group to pay for the local loops for the out group. The amount of subsidy determined the price of service for the out group and thereby defined the size of the in group. Within this context, the universal suffrage metaphor played an important role by providing justification for public policy intervention as it suggested that access to telephone, a critically important technology for participation in the life of the community, was almost a right of every citizen. An elaborate system of subsidies established by the regulators eventually expanded the telephone service to everybody.

The extension of telephone service to almost the entire population settled the question of universal telephone service until the recent convergence of telephone and computer technologies transformed the telephone network from a vehicle for the transmission of voice telephony signals to a platform for the delivery of a whole host of information services such as 911

---

[10]Although we did not come across any historical evidence explicitly linking universal suffrage and universal telephone service, it is quite apparent that the principle of universalism was at the root of both. For example, Assemblywoman Gwen Moore, during her universal service campaign in California, made this connection when she said "[I]f the freedom to communicate is a fundamental right then access to the means of communication must also be a fundamental right. Without access, one cannot be a part of the telecommunicating community" (Jacobson, 1989, p. 59). Thus we see that the universalist ideal created the conceptual structures within which all the "universal" initiatives were taken.

[11]According to Littlechild (1979), network externalities have been recognized as a factor in telephony right from the 1890s. The generation of a consumer surplus due to larger network size is fairly obvious, although there is presently little agreement on the pricing implications of this positive externality. See Taylor (1994) and Mitchell and Vogelsang (1991) for rigorous economic analyses of network externality.

emergency service, home banking, distance learning, remote medical diagnostics, surveillance, energy management, special services for the hearing impaired, automatic language translation, voice mail, computer conferencing, access to databases, and others. This proliferation of telephone-based services has changed the issue of universal telephone service from "one household, one telephone" to "one household, how many services?" There is general agreement that the definition of universal service needs to be extended beyond basic voice communication (Gillan 1986; Hadden, 1991a, 1991b; Hudson, 1994; Information Infrastructure Task Force, 1993; National Telecommunications and Information Administration [NTIA], 1988, 1991; O'Connor, 1991; Office of Technology Assessment [OTA], 1990; Parker, Hudson, Dillman, & Roscoe, 1989; Williams, 1991). However, nobody has been able to develop universally accepted criteria for including some services in the universal service package and excluding others. Within this changed context, the universal suffrage metaphor, which had a definite closure (one person, one vote) is unable to tackle the open-endedness of a multiple-services environment. Quite obviously, there is a need for a new metaphor.

The universal education metaphor seems more appropriate because it also has an open-endedness to it that is akin to modern telecommunications. The question of universal education never had closure as it dealt with a continually evolving phenomenon. The definition of universal education gradually expanded from elementary school to middle school to high school and even touched on higher education. How were universal education policies developed for a "service" that had no well-defined boundaries? This experience with universal education perhaps has much to offer our current efforts to develop universal service policies for advanced telecommunications services.

IMPLICATIONS FOR UNIVERSAL SERVICE IN ADVANCED TELECOMMUNICATIONS

This concluding section draws on the experiential gestalt of our previous efforts in universal education to sketch out implications for the development of universal service in advanced telecommunications. Based on our analysis in this chapter, we have distilled the following three broad principles that have a bearing on our current efforts:

An a priori definition of universal service is not a necessary condition for the development of universal service in advanced telecommunications. Our fixation on defining universal service has led us to focus our energies almost entirely on developing a new definition. The assumption here is that once a new definition is developed, the rest of the decision-making process will be relatively simple. This logic has such a strong grip on our imagination that repeated failures have not deterred us from our efforts to develop a new definition for universal service. The history of universal education beckons us to rethink our assumptions.

The development of universal education continued even in the absence of an adequate definition for a basic concept like the "public school." Although it may seem trivial in retrospect, a considerable amount of energy was expended in the search for such a definition in which legislatures, religious groups, and even the courts were involved. The debate was especially contentious because designation as a public school made an institution eligible for public finance, but also subjected it to state supervision over instructional content (Good, 1956). However, progress in universal education did not await the resolution of these conflicts. It progressed in spite of, and possibly because of, this definitional ambiguity.

Initiatives in universal higher education were undertaken in spite of no clear answers for many fundamental questions: What exactly is higher education? Should its definition be restrictive and limited to 4-year colleges, or could it be expansive and include community colleges, technical institutes, and all post-high school vocational programs? What are the benefits and costs of higher education? There was at best an expectation that universal higher education would provide benefits in the form of enhanced productivity and an enriched democratic process. It was, however, difficult to quantify these benefits (Bowles, 1966; Hansen & Witmer, 1972). Even an understanding of the costs involved was hazy at best as estimates of the percentage of population who would benefit from higher education ranged from 25% to 49%, creating uncertainty about the size of the investment required (Miller, 1971). Universal higher education grew in an environment within which higher education was "unable to define with precision its purposes, to measure with clarity its processes, or to quantify with certainty its outputs" (Olivas, 1979, p. 1). It was within this overall context of great ambiguity that the final and perhaps the most important question—who should have access to higher education—was tackled.

The parallels with universal telecommunications service are striking. We are now faced with a similar set of questions: What should be the definition of universal service? What will be its benefits and costs? Who should have access to advanced telecommunications services? The experience with universal education suggests that our current efforts to define universal service is a misdirected effort. Perhaps it will never be possible to develop a national consensus on a new definition. Furthermore, the lack of definitive answers to these questions is not necessarily a roadblock for the development of universal service.

Exemplars in the form of implemented programs have a much greater influence on the nature and scope of universal service than formal definitions. The concept of universal service is inherently an ambiguous idea. It springs from notions of equality and participatory democracy. The idea resonates with high ideals but has a vagueness that makes its translation into implementable programs problematic. However, precisely because of this dilemma, tentative real-world solutions have a much greater impact on the further development of the idea than theoretical formulations. Even if formally imprecise, these "solutions" have a metaphoric impact that makes the practical heroic and thereby bridges the gap between the mundane and the ideal. These solutions become the exemplars against which all subsequent efforts are measured. Like Massachusetts and Virginia for early education, and the California Higher Education System and the City University of New York for higher education, exemplars become the surrogate for the ideal that is difficult to conceptualize. The current debate about universal service is still in the preexemplar stage. The policies that encourage decentralization and allow the different states to develop their own solutions will facilitate the emergence of exemplars. They will draw attention away from the search for definitions, to an examination of the pros and cons of different exemplars, and the development of hybrids suitable for each state.

Universal service policy is in reality a bundle of policies whose elements arise from different sources in an synchronous manner and are later harmonized into an integrated policy framework with local variations in each jurisdiction. We tend to view universal service as a monolithic concept that can be realized by a masterful grand plan. The development of universal education reveals that in reality the process is far more diffuse and disorganized. The policy framework that sustains universal education is highly modular: it has in fact been described as a "complex network of interrelated rules" (Katz, 1976). These modules develop separately and are later organized into integrated systems. There is considerable variance in the makeup and composition of each module, and they, in turn, can be arranged in different combinations to create a wide variety of systems. This diversity is reflected in the great variation in the education systems of different states, all of whom offer universal education in their own peculiar ways. This experience with universal education suggests that a similar modularized approach would be more appropriate for universal service in advanced telecommunications. Instead of working on an overarching framework, we should create a decentralized policy environment that would facilitate the development of a wide variety of policy modules—eligibility criteria, service packages, finance, and others—that could later be organized into integrated systems.

ACKNOWLEDGMENTS

Correspondence concerning this chapter should be addressed to Harmeet Sawhney, Department of Telecommunications, Indiana University,

Bloomington, IN 47405, while electronic mail may be sent via Internet to: hsawhney@indiana.edu, or to Krishna Jayakar, Department of Telecommunications, Indiana University, Bloomington, IN 47405. E-mail: kjayakar@indiana.edu.

REFERENCES

Alexander, K., & Jordan, K. F. (1973). *Legal aspects of educational choice: Compulsory attendance and student assignment*. Topeka, KS: National Organization of Legal Problems in Education.

Arboleda, J. (1981). *Open admissions to higher education and the life chances of lower class students: A case study from Columbia*. Unpublished doctoral dissertation, Indiana University, Indiana.

Birenbaum, W. M. (1971). Something for everybody is not enough. In W. T. Furniss (Ed.), *Higher education for everybody?* (pp. 65–82). Washington, DC: American Council on Education.

Black, M. (1962). *Models and metaphors: Studies in language and philosophy*. Ithaca, NY: Cornell University Press.

Bowles, F. (1966). Observations and comments. In E. J. McGrath (Ed.), *Universal higher education* (pp. 235–245). New York: McGraw-Hill.

Brock, G. (1981). *The telecommunications industry: The dynamics of market structure*. Cambridge, MA: Harvard University Press.

Brooks, J. (1976). *Telephone: The first hundred years*. New York: Harper & Row.

Brown, R. H. (1976). Social theory as metaphor: On the logic of discovery for the sciences of conduct. *Theory and Society, 3*, 169–197.

Educational Policies Commission. (1964). *Universal opportunity for education beyond the high school*. Washington, DC: National Education Association of the United States.

Edwards, N., & Richey, H. G. (1963). *The school in the American social order*. Boston: Houghton Mifflin.

Fagen, M. D. (Ed.). (1975). *History of engineering and science in the Bell system: Vol. 1. The early years, 1876–1925*. Warren, NJ: Bell Telephone Laboratories.

Fischer, C. (1987). The revolution in rural telephony: 1900–1920. *Journal of Social History, 21*, 5–26.

Fischer, C. (1992). *America calling: A social history of the telephone to 1940*. Berkeley, CA: University of California Press.

Fiumara, G. C. (1995). *The metaphoric process: Connections between language and life*. New York: Routledge.

Garnet, R. (1985). *The telephone enterprise*. Baltimore: Johns Hopkins University Press.

Gentner, D., & Jeziorski, M. (1993). The shift from metaphor to analogy in western science. In A. Ortony (Ed.), *Metaphor and thought* (pp. 447–480). New York: Cambridge University Press.

Gillan, J. (1986). Universal telephone service and competition: The rural scene. *Public Utilities Fortnightly, 117*, 21–26.

Good, H. G. (1956). *A history of American education*. New York: Macmillan.

Hadden, S. (1991a). *Regulating content as universal service* (Working Paper, Policy Research Project: "Universal Service for the Twenty-First Century"). Austin, TX: The University of Texas at Austin, Lyndon B. Johnson School of Public Affairs.

Hadden, S. (1991b). Technologies of universal service. In The Institute for Information Studies (Ed.), *Annual review, 1991: Universal telephone service: Ready for the 21st Century?* (pp. 53–92). Nashville, TN: The Institute for Information Studies.

Hansen, W. L., & Witmer, D. R. (1972). Economic benefits of universal higher education. In L. Wilson & O. Mills (Eds.), *Universal higher education: Costs, benefits, options* (pp. 19–39). Washington, DC: American Council on Education.

Hillesheim, J. W., & Merril, G. D. (1971). *Theory and practice in the history of American education: A book of readings*. Pacific Palisades, CA: Goodyear Publishing.

Hudson, H. (1994). Universal service in the information age. *Telecommunications Policy, 18*, 658–667.

Information Infrastructure Task Force. (1993). *The national information infrastructure: Agenda for action*. Washington, DC: National Telecommunications and Information Administration, National Information Infrastructure Office.

Jacobson, R. (1989). *An open approach to information policy making: A case study of the Moore Universal Telephone Service Act*. Norwood, NJ: Ablex.

Jaffe, A. J., & Adams, W. (1972). Two models of open enrollment. In L. Wilson & O. Mills (Eds.), *Universal higher education: Costs, benefits, options* (pp. 223–251). Washington, DC: American Council on Education.

Johnson, M. (1981). Introduction: Metaphor in the philosophical tradition. In M. Johnson (Ed.), *Philosophical perspectives on metaphor* (pp. 3–47). Minneapolis, MN: University of Minnesota Press.

Karabel, J. (1972). Perspectives on open admissions. In L. Wilson & O. Mills (Eds.), *Universal higher education: Costs, benefits, options* (pp. 265–286). Washington, DC: American Council on Education.

Katz, M. S. (1976). *A history of compulsory education laws*. Bloomington, IN: Phi Delta Kappa Educational Foundation.

Kotin, L., & Aikman, W. F. (1980). *Legal foundations of compulsory school attendance*. Port Washington, NY: Kennikat Press.

Lakoff, G., & Johnson, M. (1980). *Metaphors we live by*. Chicago, IL: University of Chicago Press.

Lavin, D. E., Alba, R. D., & Silberstein, R. A. (1981). *Right versus privilege: The open-admissions experiment at the City University of New York*. New York: Free Press.

Littlechild, S. C. (1979). *Elements of telecommunications economics*. New York: Peter Peregrinus.

Miller, J. L., Jr. (1971). Who needs higher education? In W. T. Furniss (Ed.), *Higher education for everybody?* (pp. 94–105). Washington, DC: American Council on Education.

Mitchell, B. M., & Vogelsang, I. (1991). *Telecommunications pricing: Theory and practice*. New York: Cambridge University Press.

Mueller, M. (1989). The switchboard problem: Scale, signaling, and organization in manual telephone switching, 1877–1897. *Technology & Culture, 30*, 534–560.

Mueller, M. (1993). Universal service in telephone history: A reconstruction. *Telecommunications Policy, 17*, 352–369.

Munday, L. A., & Rever, P. R. (1971). In P. R. Rever (Ed.), *Open admissions and equal access* (pp. 75–96). Iowa City, IA: The American College Testing Program.

National Telecommunications and Information Administration. (1988). *Telecom 2000: Charting the course for a new century* (NTIA Special Publication 88-21). Washington, DC: U.S. Government Printing Office.

National Telecommunications and Information Administration. (1991). *The NTIA infrastructure report: Telecommunications in the age of information* (NTIA Special Publication 91-26). Washington, DC: U.S. Government Printing Office.

Nuessel, F. (1988). Metaphor and cognition: A review essay. In M. Danesi (Ed.), *Metaphor, communication, and cognition* (pp. 9–22). Toronto, Canada: Toronto Semiotic Circle.

O'Connor, B. (1991). Universal service and NREN. In The Institute for Information Studies (Ed.), *Annual Review, 1991: Universal telephone service: Ready for the 21st Century?* (pp. 93–140). Nashville, TN: The Institute for Information Studies.

Office of Technology Assessment. (1990). *Critical connections: Communications for the future* (OTA-CIT-407). Washington, DC: U.S. Government Printing Office.

Olivas, M. (1979). *The dilemma of access: Minorities in two year colleges*. Washington, DC: Howard University Press.

Parker, E. B., Hudson, H. E., Dillman, D. A., & Roscoe, A. D. (1989). *Rural America in the information age: Telecommunications policy for rural development*. Boston, MA: University Press of America.

Porter, K. H. (1969). *A history of suffrage in the United States*. New York: Greenwood. (Original work published 1918.)

President's Commission on Higher Education. (1947). *Higher education for American democracy, Vol. 1*. Washington, DC: U.S. Government Printing Office.

Rossman, J. E., Astin, H. S., Astin, A. W., & el-Khawas, E. H. (1975). *Open admissions at City University of New York: An analysis of the first year.* Englewood Cliffs, NJ: Prentice-Hall.

Sawhney, H. (1994). Universal service: Prosaic motives and great ideals. *Journal of Broadcasting & Electronic Media, 38*, 375–395.

Schrag, P. (1971). Open admissions to what? In P. R. Rever (Ed.), *Open admissions and equal access* (pp. 49–53). Iowa City, IA: The American College Testing Program.

Taylor, L. D. (1994). *Telecommunications demand in theory and practice.* Boston: Kluwer.

Thompson, R. L. (1972). *Wiring a continent: The history of the telegraph industry in the United States, 1832–1866.* New York: Arno Press. (Original work published 1947.)

Urban, W., & Wagoner, J. (1996). *American education: A history.* New York: McGraw-Hill.

U.S. Office of Education. (1971). *Report on higher education* (OE. 50065). Washington, DC: U.S. Government Printing Office.

Williams, F. (1991). *The new telecommunications: Infrastructure for the information age*. New York: The Free Press.

Williamson, C. (1960). *American suffrage: From property to democracy, 1760–1860*. Princeton, NJ: Princeton University Press.

Willingham, W. W. (1970). *Free-access higher education*. New York: College Entrance Examination Board.

Wisconsin v. Yoder. (1972). 406 U.S. 205

3

Unilateral and Bilateral Rules: A Framework for Increasing Competition While Meeting Universal Service Goals in Telecommunications

Barbara A. Cherry
Steven S. Wildman
Northwestern University

The telecommunications industry in the United States and much of the rest of the world is in a period of transition from monopoly to competition. Yet, movement toward a competitive industry does not imply a total absence of regulation. Society still expects telecommunications providers to make contributions toward the achievement of policy goals that would not be made if the telecommunications industry were unconstrained by regulatory and legal requirements. These expectations are exemplified by various performance obligations, such as ubiquity of service and affordable prices, that are critical elements of universal service policy in the United States and most other industrialized countries.

Different approaches will be required to achieve these policy goals in a more competitive industry than those that worked with traditional regulated monopolies and postal telephone and telegraph entities (PTTs). In recognition of this fact, governments have been revising the rules governing providers—the Telecommunications Act of 1996 in the United States being the most visible recent example. Unfortunately, these deregulatory initiatives are being advanced without a clear vision of how they interrelate with each other or how they might be coupled structurally to accomplish various goals.

This chapter develops a typology for mapping social goals regarding marketplace activities to the types of regulatory interventions that are needed to accomplish those goals. Fundamental to this typology and its application is the distinction between goals that can be achieved through requirements unilaterally imposed by government on economic actors and goals that can be realized only through the use of bilateral arrangements in which some form of compensation or privilege is provided by government to elicit the performance of activities that would otherwise be unremunerative. Furthermore, the nature of the required bilateral arrangements will also vary, depending, in particular, on the degree of vulnerability to regulatory expropriation of investment.

To date, legislators and regulators have not recognized the distinction between unilateral and bilateral rules or its implications for the design of regulations for telecommunications industries. This is not surprising because telecommunications services historically have been provided by either state-franchised or state-owned monopolies. In both cases, the state's vested interest in ensuring that providers had the financial wherewithal to provide requested services made moot a distinction between unilateral and bilateral arrangements. However, as we come to rely on the performance of competitive providers to accomplish policy objectives, it is critical that telecommunications policies be crafted with a more sophisticated understanding of the extent to which various types of interventions are compatible with the workings of competitive markets and the degree to which departures from the ideal of an unconstrained marketplace may be required for the accomplishment of important policy goals. The typology of policy goals and interventions presented in this chapter was developed to provide the conceptual underpinnings for these types of assessments.

A FRAMEWORK FOR ENSURING COMPATIBILITY BETWEEN POLICY GOALS AND POLICY INTERVENTIONS IN A MORE COMPETITIVE TELECOMMUNICATIONS INDUSTRY

Some social goals are not achievable in an unregulated marketplace for a variety of reasons. Society may not approve of some products supplied by markets, markets may suffer from various imperfections leading to inefficiency in the supply of goods and services society does want, or private markets may not serve some individuals who society would like to have served. Policy responses to these problems may take a variety of forms, but, to be effective, such responses must satisfy certain compatibility criteria.

Compatibility Between Social Goals and Regulatory Policies

Any plan for a more competitive telecommunications industry must have both a *long-term* vision that defines policy goals and appropriately matches them with regulatory instruments to achieve those goals, and mechanisms for dealing with the *transition* from the current state of affairs to the one that is desired in the long term. In both cases, two types of compatibility are prerequisite to success. One is that each social goal pursued be achievable by the regulatory intervention employed on its behalf. The second is that any particular combination of social goals be achievable given the aggregate set of interventions employed in the pursuit of each individually. That is, it is important that policy interventions, which might seem appropriate in isolation, not conflict with each other to the extent that, taken in combination, they preclude the accomplishment of one or more of the goals pursued.

There are a number of reasons why either form of compatibility may not be realized. An individual goal-intervention combination may not be compatible because the intervention does not address critical problems associated with achieving the goal. For example, subsidized prices for local calling will not increase telephone subscribership among households who refuse to take service due to high toll bills. Goals may also be inherently incompatible with each other, which precludes their joint realization. Fiber-optic cable to the home and low-cost local service are examples of goals that cannot be achieved simultaneously, at least not with current technology. The threat to compatibility is the possibility that certain types of regulatory policies developed to promote universal service while telecommunications markets are being opened to competition may render the affected carriers financially unable or unwilling to provide service and preclude their contributing to the attainment of universal service policy goals.

Principles to Ensure Goal-Policy Compatibility

Unilateral and Bilateral Rules. Regulation may take an almost infinite variety of forms. Government may supply a service or product, as with the provision of telecommunications services by government-owned PTTs in many European countries, or government may regulate privately owned suppliers, as with the federal and state governments' regulation of telecommunications carriers in the United States. If we exclude direct supply of a product or service by government, all forms of regulation may be classified as either *unilateral rules* or *bilateral rules.*

Unilateral rules are performance requirements imposed by government on firms as a condition for providing service without any assurance by government that the affected firms will be able to generate revenues sufficient to cover the associated costs. Minimum wage laws, workplace safety requirements, and product reliability and safety standards are examples of the many unilateral requirements that affect economic activities. Some unilateral rules may also be viewed as granting a benefit or privilege by government. An example is a tax credit, which can also be viewed as a change in a performance requirement. This chapter is concerned with unilateral rules that impose performance requirements rather than grant benefits.

Bilateral rules differ from unilateral rules in that with bilateral rules firms receive some form of compensation or special consideration in exchange for meeting government-specified performance obligations. With a bilateral rule, the government and a regulated firm have mutual and specific obligations toward each other. Furthermore, although bilateral rules are typically accepted by the affected firms, unilateral rules are simply imposed by government as a condition for doing business.

Within the category of bilateral rules, we define two types. *Bilateral agreements* are government-specified performance requirements that are coupled with financial compensation for costs associated with meeting the requirements. Other than providing prespecified levels of compensation, the government assumes no responsibility for the financial health of firms that are parties to bilateral agreements. The government is, in effect, buying a service from these firms, and their ability to earn a fair return on their efforts is their own concern. *Bilateral commitments* are performance obligations accepted by firms in exchange for which the government accepts some degree of responsibility and provides some form of assurance for the financial health of the firms taking on these requirements, including safeguards against the threat of regulatory expropriation of the investments required to provide service.

Lifeline and Link-up programs in the United States, which provide funding to local exchange carriers (LECs) for the provision of service to low-income customers, are examples of bilateral agreements. Although the government provides explicit payments to LECs for providing these services, the carriers accept these contracts independent of their other regulatory obligations, and they alone are responsible for whatever profits or losses they realize providing them.

The monopoly franchises that in the past were traditionally granted to public utilities in the United States—described as regulatory contracts by Goldberg (1976, 1980)—are a form of bilateral commitment. As Goldberg and others (Wiggins, 1991; Williamson, 1976) describe the regulatory contract, the regulated firm agrees to make substantial sunk investments and provide service subject to restrictions on price and the nature of the services provided that would not satisfy normal risk-return criteria without some assurance by government of a reasonable chance of earning a fair return on these investments.

Under franchise monopoly, the assurance provided by government took the form of restrictions on competitive entry and the understanding that regulated firms would be allowed to recover in their rates all prudently incurred costs. Patent laws are another example of a bilateral commitment involving a restriction on entry subsequent to the incurrence of sunk costs. Here the prospect of earning a fair return on investments in innovations is increased through grant of a patent, which provides for exclusive use of the innovation covered by a patent for a substantial period of time, in exchange for public disclosure of the invention in the patent application.

PTT provision of telecommunications services is an alternative response to the financial vulnerability associated with large sunk costs, and, as such, exhibits similar characteristics. Whether the sunk costs are committed by franchised private providers or by a government-owned PTT on behalf of ratepayers as owners, assurance of a continued right to provide service at compensatory rates is still required to ensure a reasonable return on the resources committed. Thus, both PTTs and franchised private providers were set up as monopolists with customer bases protected from entry.

There are limits, however, on governments' use of unilateral and bilateral rules. Important restrictions are imposed by provisions of both the federal and state constitutions in the United States and by related restrictions in the laws of other countries. The reasons for these restrictions on state actions are discussed later. The typology of economic regulation just discussed is depicted in Fig. 3.1.

Competition and Choices Among Rules. There are three distinct situations where policy goals may not be achieved through unilateral rules imposed on competitive firms. First, only one or a few firms may be able to generate revenues sufficient to cover the costs of the unilateral requirements. If too many firms exit, the industry will no longer be competitive; and, in the extreme, no firms will be able to offer service.

Second, the costs of a rule may not be shared equally by all firms, either because the rule is asymmetrically imposed or because firms differ in their ability to evade it. If the financial burden of a unilateral rule is greater for some firms than for others, those firms bearing the greatest burden will be driven from a competitive industry unless, of course, they benefit from off-setting advantages of another type. This is a problem if those for whom the burden is least are better able to avoid obeying the rule, or if differences in the incidence of burden reflect asymmetries in enforcement or regulatory design that have nothing to do with the relative merits of the activities of different firms. In this case, otherwise efficient firms may be driven from an industry and competitive outcomes will be characterized by adverse selection favoring those firms that are best at either evading the unilateral rule or influencing the political process to their advantage.

In many situations where unilateral rules are not compatible with competition, whether due to the high costs of compliance or to asymmetry of incidence, it may be possible to preserve competition through compensation provided to firms under a bilateral agreement. Food stamps given to low-income individuals in the United States is an example. If grocery stores were required to sell food below cost to low-income consumers, competition would favor the stores most successful in discouraging their patronage. Use of food stamps to compensate grocery stores for such sales solves this problem. Proposals for low-income vouchers for telecommunications services rely on a similar logic, although the analogy may not be valid in all circumstances (see, e.g., Panzar & Wildman, 1993, 1995).

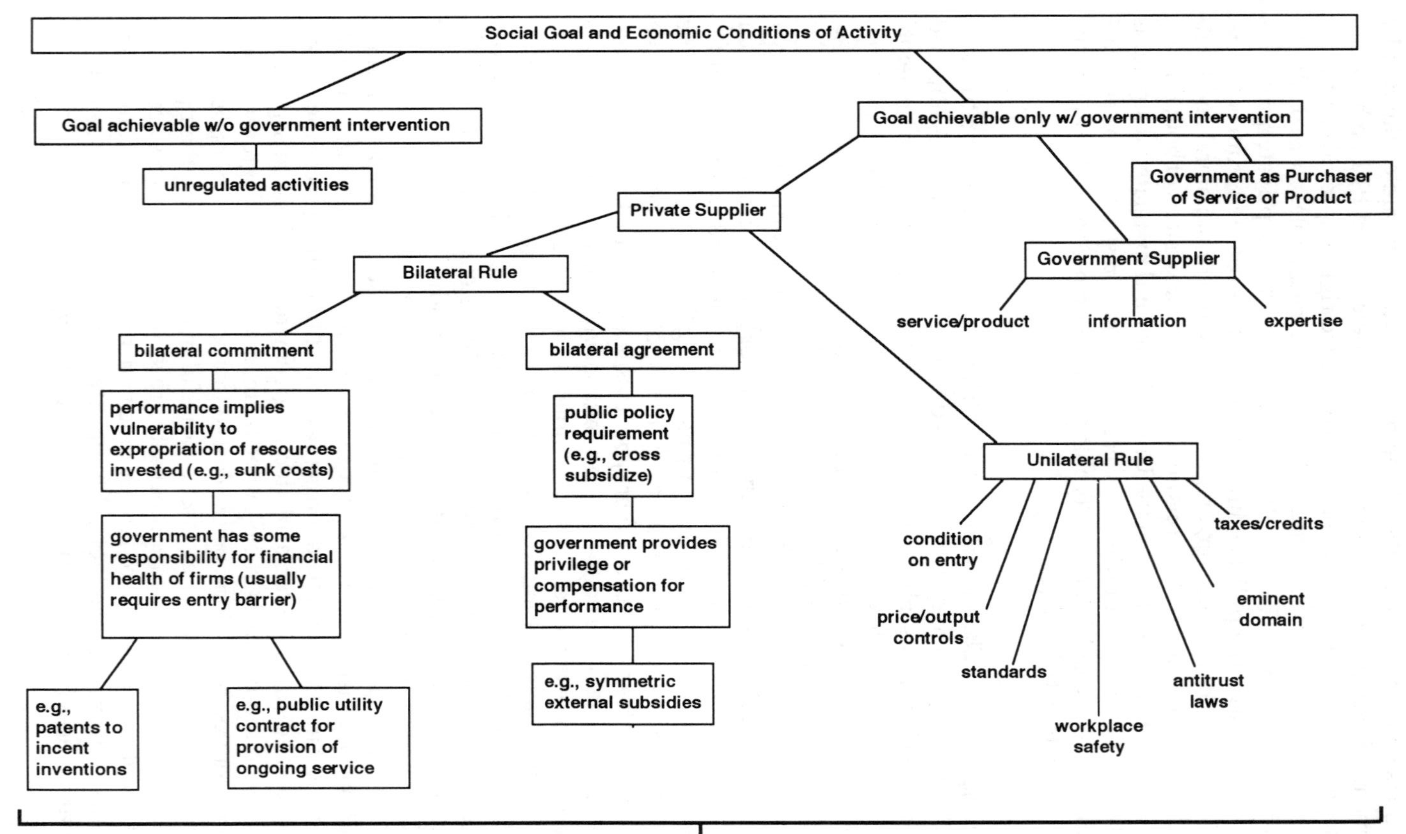

FIG. 3.1 Social goal and economic conditions of activity.

Third, the desired behavior may be financially feasible only if competition is suspended. A bilateral commitment is then required, as with the regulatory contract situation described earlier. As Goldberg (1976, 1980) pointed out, when the supply of a product or service is characterized by substantial sunk costs, the risk that customers may turn to an alternative supplier after sunk costs have been incurred increases the price at which a firm will be willing to offer service and may actually preclude the provision of service entirely. In this situation, a bilateral commitment that precludes customer purchases from competing providers, or perhaps specifies compensation to the original supplier in such an eventuality, reduces supplier risks and the price at which service can be offered. Note, however, that the barrier to entry is effective after the firm has accepted a request to provide service. There may still be vigorous competition ex ante among firms vying to be the service provider, as in auction or bidding situations.

APPLYING THE FRAMEWORK TO UNIVERSAL SERVICE POLICY

Traditional Universal Service Policy Under Monopoly

Although there may be disagreement as to the historical meaning of the term (Mueller, 1993), universal service has generally come to mean access to basic, analog voice grade service at affordable rates. An important reason for making universal service a fundamental goal of telecommunications policy is a belief that the social value of a ubiquitous network will not be reflected adequately in individual consumers' decisions to take service. The value of a network to each of its subscribers grows as the numbers increase, and this positive externality will not be fully internalized if there is no mechanism by which existing subscribers can make the contributions required to get new subscribers to join who would not do so without some source of support. Also underlying the concept of universal service is the notion that equity and fairness require that access be affordable to all individuals.

But universal service has also come to mean more than just affordability and ubiquity. It is also identified with a maze of regulatory mechanisms that have created a price structure that, although addressing various universal service policy subgoals, would not have developed in competitive markets. For example, in the United States the prices of various telecommunications services are geographically averaged and often include rate elements designed to collect funds to help defray the costs of service for special classes of subscribers, such as low-income households, hearing-impaired individuals, or individuals served by high-cost local exchanges (Weinhaus et al., 1992, 1993). This price structure has been maintained traditionally through grants of monopoly franchises. Similarly, in Europe, universal service has been associated with a noncompetitive price structure maintained through the monopoly provision of service by either government-owned or government-designated enterprises.

Table 3.1 lists some of these universal service subgoals and their corresponding social and economic problems. Table 3.1 also lists the policies employed traditionally in pursuit of these subgoals under franchise monopoly in the United States. With the exception of dual-party relay services for the hearing-impaired, which are provided under a separate bilateral commitment, all of these policies were administered as components of a bilateral commitment based on franchise monopoly.

Substantial changes have occurred in U.S. federal and state telecommunications regulation in recent years as restrictions on entry have been re-

TABLE 3.1

Universal Service Subgoals and Regulations Under Franchise Monopoly

Social Goal	*Economic and Social Problems*	*Traditional Policies Under Franchise Monopoly*
1. Provide telecommunications services at reasonable prices.	1. Market power results in high prices for an "essential" service and reduction in network penetration to less than socially efficient levels.	1. Common carrier obligations, price or earnings regulation, and inter- and intraservice support flows.
2. Provide economically disadvantaged individuals with access to certain basic telecommunications services.	2. Individuals who cannot afford cost-based prices value service at less than the social value of their subscription.	2. Lifeline (subsidized monthly) rates; linkup programs to subsidize installation fees.
3. Provide individuals with disabilities access to special basic telecommunications services.	3. The costs of services needed for those with certain disabilities exceed "fair" prices, and, if reflected in prices, may reduce penetration to less than socially efficient levels.	3. Dual-party relay service is provided below cost to those with disabilities. Providers bid to provide service; government selects the provider and provides for funding.
4. Provide individuals who are costly to serve with access to certain basic telecommunications services and capabilities.	4. Locational factors raise the cost of service to some people above "fair" prices; cost-based prices may reduce penetration below socially efficient levels.	4. Carrier of last resort obligations, interservice and intraservice customer class support flows, and high cost assistance funding.
5. Provide telecommunications services at some minimum level of quality.	5. Individual providers fail to fully internalize costs and benefits of service quality improvements.	5. Certification requirements on providers; service quality regulations; and interconnection-related requirements.

laxed and competition encouraged. This pattern is being repeated in other countries liberalizing their telecommunications industries. With increasing entry into local markets, questions concerning the rules and responsibilities applied to new entrants have and will continue to be a major source of contention.

In the United States, new entrants generally do not bear the same regulatory obligations as incumbents, giving rise to asymmetric regulatory burdens. Such asymmetric burdens cannot be sustained in a competitive equilibrium, and are recognized by both scholars (Shankerman, 1994) and government officials (Haring, 1984) as a threat to policies promoting universal service. Notwithstanding these sustainability problems, most analyses of local competition and asymmetric regulation have concentrated on their economic efficiency properties. For example, Weisman (1994) discussed technical efficiency losses and dynamic efficiency losses as well as the breeding of inferior competitors who are adept at imitation rather than innovation. The typology of unilateral and bilateral rules presented here provides a framework for examining the long-term and transitionary sustainability problems associated with recent universal service policies and for designing the universal service policies of the future.

Competition and Sustainable, Long-Term Universal Service Policies

The general pattern of telecommunications deregulation in the United States has been to relax restraints on competitive entry while reducing, but not eliminating, restraints on prices and earnings for incumbents. At the same time, regulators have attempted to retain most of the performance obligations listed in Table 3.1 as unilateral requirements, typically with higher performance expectations for incumbent providers. As a result, almost all of the old bilateral arrangements have been converted to asymmetrically imposed unilateral rules.

Use of Unilateral Rules. For unilateral rules to be features of a long-run competitive equilibrium, they must be applied symmetrically. If not, the asymmetrically advantaged firms will drive out the rest. Some of the current set of unilateral rules pose no problems in the long run because they will, of necessity, simply disappear. Examples include price and earnings regulation because competition will force prices to accurately reflect underlying costs. However, the various cross-subsidies embedded in current prices reflect a social unwillingness to accept the outcomes of cost-based pricing. As discussed in Cherry and Wildman (1996), such cross subsidies cannot be maintained as unilateral rules in a competitive telecommunications industry because enforcement of symmetry would be extremely difficult and the incentive to avoid these obligations would be great.

To ensure symmetry in cross-subsidy burdens would require that each competing provider's share of cross-subsidy recipients be proportional to its market share, that the cost characteristics of the subsidy recipients assigned to each provider be similar, and that the pool of subsidy recipients be continuously reallocated as market shares change and as firms enter and exit the market. Considerably more cost information than is currently available would be required to ensure that some providers did not get more than their fair share of the highest cost subsidy recipients—information that is likely to be costly to acquire and for which regulators would have to rely on cooperation by the affected firms. Furthermore, to date local service competition has taken the form of entrants targeting considerably narrower geographic areas and sets of customers than those served by incumbents. By carefully selecting the regions and customer classes they serve, entrants can reduce their cross-subsidy burdens relative to incumbents. As long as incumbents are required to offer geographically averaged rates, as they are in the United States, the ability to target their service offerings is likely to be a substantial competitive advantage to their competitors.

Common carrier obligations, which require carriers to provide service to similarly situated customers on equivalent terms, are also likely to be unsustainable with competition (Noam, 1994). Although such obligations may be easy to enforce for monopolists, it would be virtually impossible to police the marketing plans of multiple providers to ensure that information of competitive offerings is not selectively targeted.

Carrier of last resort obligations require a provider to expand capacity to serve new customers on reasonable request and constitute a barrier to exit from areas already served. Because the very need to enforce carrier of last resort obligations implies that service must be provided at noncompensatory rates, such obligations also pose the threat of burdensome cross-subsidies that carriers will have strong incentives to avoid when possible. Because new entrants can assure themselves lower burdens than incumbents by carefully selecting their service territories, unless symmetry in geographic coverage is required, asymmetric incidence of carrier of last resort obligations is virtually assured.

Use of Bilateral Rules. The extent to which sustainability can be achieved by converting unilateral rules to bilateral agreements depends on whether competition would naturally emerge in the absence of asymmetric obligations. Subsidies for some classes of customers may be handled through bilateral agreements if the group of subsidy recipients is sufficiently small that the incremental cost of the facilities required to serve such recipients is low, especially if contracts with providers are of short duration and growing demand was expected to create a need for similar facilities in the near future. But, if the class of subsidy recipients constitutes a substantial portion of all customers, the sunk cost investments at hazard to regulatory expropriation will also be large, and assurances of the type that can be provided only through a bilateral commitment are likely to be required. Thus, policy decisions as to the size of the class of subsidy recipients will affect the extent to which open competition is actually achievable.

As stated earlier, common carrier and carrier of last resort obligations are not likely to be sustainable as part of a competitive equilibrium if imposed by unilateral rules. To the extent the incremental sunk costs required to meet common carrier obligations are small, bilateral agreements should be sufficient; if they are large, bilateral commitments will be required.[1] As for carrier of last resort obligations, under current technology local loops still represent substantial sunk investments that are vulnerable to expropriation—either through less than compensatory rates or due to asymmetric sharing of the burden of service for higher cost customers. Given these vulnerabilities, it seems likely that bilateral commitments will be necessary for some time to come if carrier of last resort obligations are to be maintained as a component of universal service policy.

These vulnerability problems may be substantially overcome if technology evolves so that carriers find it in their interest to install substantial redundant capacity. If such naturally occurring excess capacity can be employed in the service of support recipients, the sunk cost vulnerabilities of providing service to these subscribers may be significantly reduced. One of the features of competition in toll services has been the installation of enough capacity that, even if a major carrier were to exit the market, the remaining competitors would still have sufficient capacity to serve the entire market. Continued development of wireless technology may lead to an analogous situation for local services in the future. However, the incentive to avoid an obligation to provide service at unremunerative rates would remain.

The problems likely to arise if current universal service policies are continued as local telecommunications markets become more competitive, along with remedies for these problems, are summarized in Table 3.2.

Managing the Transition to a More Competitive Industry

Because bilateral commitments are likely to be an important component of telecommunications regulation for some time to come, the credibility of promises by the government as a party to bilateral commitments with regu-

[1]The incentive to cheat on common carrier obligations through price discrimination would still be present under a bilateral agreement. One way to counter this tendency would be to set the penalty for cheating high enough to serve as a deterrent.Under a bilateral commitment, the fear of being denied a continued right to serve and thereby the ability to recover substantial sunk costs could promote compliance. However, once such a penalty is invoked, compliance is likely to deteriorate further until these obligations are fully shifted to another carrier, which is contrary to the public interest. This suggests that this most severe penalty should be invoked only for extreme violations.

TABLE 3.2

Unsustainability of Unilateral Rules and Proposed Remedies in a More Competitive Industry

Current Policy	*Sustainability Problem*	*Remedy*
Asymmetric unilateral rules in general.	Asymmetrically disadvantaged firms will exit industry.	Apply unilateral rules symmetrically.
Cross subsidies as symmetric unilateral requirements.	Symmetry not enforceable.	Replace with bilateral agreement if sunk cost exposure is not large. Otherwise, replace with bilateral commitment.
Common carrier obligations as symmetric unilateral rules.	Symmetry not enforceable.	Replace with bilateral agreement if sunk cost exposure is not large. Otherwise, replace with bilateral commitment.
Support for low-income customers as symmetric unilateral rules.	Symmetry probably not enforceable.	Bilateral agreements should work.
Carrier of last-resort obligations and service requirements for high-cost areas are symmetric unilateral rules.	Conditions for symmetric application not yet satisfied and enforceability doubtful.	Bilateral commitment is needed until technology changes.

lated firms will be important to its ability to get service providers to participate in policy initiatives in the future. For practical reasons independent of any moral obligations, government cannot casually disregard prior (often implicit) commitments to regulated firms made under the old bilateral commitment.

We believe that a promising strategy for managing a successful transition to competition is to move from reliance on entry barriers to a broader application of existing legal principles that limit government action due to preexisting circumstances. This chapter explores application of those principles as they are reflected in various provisions of the U.S. Constitution. However, it should be noted that similar provisions are typically found in the constitutions of the individual states of the United States,[2] and in the constitutions of other countries and economic confederations, such as the European Community. Table 3.3 lists the five U.S. Constitutional Clauses

[2]A more detailed treatment of these issues is presented in Cherry and Wildman (1996).

TABLE 3.3

Constitutional Limits on Government Action

Constitutional Clause	*Government Relationship to Regulated Firm*	*Government Action Subject to Limitation*	*Economic/Social Problem*	*Government Action Prohibited*	*Remedy*
Takings & Due Process Clauses	Federal/ state ↔ regulated firm	Unilateral or bilateral rule	Equity and fairness; sustainabiliy of property rights system	Confiscation	Invalidation of federal or state action; conversion of unilateral to bilateral rule by providing compensation
Supremacy (& Commerce) Clause	Federal ↔ regulated firm. State ↔ regulated firm	Conflicting unilateral or bilateral rules between state and federal governments	Equity and fairness; sustainabiliy of federal policy and of firm	Interference with federal policy; trapping of costs of firm	Invalidation of state action (i.e., federal preemption)
Contract Clause	State/private party ↔ regulated firm (at Time Period 1) Regulated firm ← state (at Time Period 2)	State impairment with preexisting private contracts or its own public contracts	Equity and fairness; sustainabiliy of property rights system	Substantial impairment of contractual obligations that is not necessary or reasonable to serve a public purpose	Invalidation of federal or state action
Ex Post Facto Clause	Federal/state → regulated firm (at Time Period 1). Regulated firm ← federal/state (at Time Period 2)	Application of new, punitive unilateral rules to prior conduct	Equity and fairness	Unfair notice of laws to citizens; laws applicable only to a particular person; abusive legislation	Invalidation of retroactive rule

that are the subjects of this analysis and summarizes the nature of their restrictions on government actions.

As indicated in Table 3.3, the relevant provisions are the Takings,[3] Due Process[4], Supremacy[5], Contract[6], and Ex Post Facto[7] Clauses of the U.S. Constitution. Although concepts of equity and fairness underlie all of these Constitutional Clauses, some Clauses specifically address different types of sustainability problems posed by the vulnerability to expropriation that is inherent in investments and other high opportunity cost commitments for which the anticipated payoff is long term and not immediate. Although judicial opinions refer typically to preexisting circumstances and investments, the economic policy problem addressed is the threat that private investments in productive activities will not be made—and societal benefits will be lost—due to fears that unanticipated changes in laws or regulations will render such investments unprofitable.

These clauses address two distinct types of sustainability problems inherent in the power of government to regulate economic activity. One relates to the role of government in establishing and guaranteeing the legal and regulatory rules for transactions among private parties. Parties to private transactions are vulnerable to changes in these rules, and the risks of adverse or favorable changes in these rules will be reflected in investment decisions. Thus an important role of the Takings, Due Process, and Contract Clauses, which place strict limits on the government's ability to change these rules, has been to ensure sufficient stability in the rules governing private transactions that private parties can make investments that promote their own and society's interests in entrepreneurial ventures with reasonable confidence.

The second type of sustainability problem relates to government's obligations and responsibilities in its contractual relationships with private en-

---

[3]The Takings Clause of the Fifth Amendment to the Constitution applies to the federal government and provides in relevant part "nor shall private property be taken for public use, without just compensation."

[4]The Due Process Clause of the Fourteenth Amendment to the Constitution applies the Takings Clause to State governments and provides in relevant part that "no person shall be ... deprived of life, liberty or property, without due process of law."

[5]The Supremacy Clause provides that "This Constitution, and the Laws of the United States which shall be made in Pursuance thereof ... shall be the supreme Law of the Land; and the Judges in every State shall be bound thereby, any Thing in the Constitution or Laws of any State to the Contrary notwithstanding." U.S. Constitution, art. VI, cl. 2.

[6]The Contract Clause provides that "No State shall ... pass any ... Law impairing the Obligations of Contracts." U.S. Constitution, art I., sec. 10, cl. 1.

[7]The Ex Post Facto Clause applicable to the federal government provides that "No bill of Attainder or ex post fact Law shall be passed." U.S. Constitution, art. I, sec. 9, cl. 3.The ex post facto prohibition applicable to the States provides that "No State shall ... pass any Bill of Attainder, [or] ex post fact Law." U.S. Constitution, art. I., sec. 10, cl. 1.

tities. Private parties transacting with government agencies are similarly vulnerable to changes in the regulatory rules and legal conditions governing these relationships. In fact, because it depends on their voluntary assent, the government's ability to contract with private parties is limited by the risks they see in such contracts. Thus, a fear that government may unilaterally breach such contracts with no remedy for private parties may foreclose the possibility of some potentially beneficial public–private arrangements. In the case of public utilities, the Takings and Due Process Clauses protect utility investors against adverse, unanticipated changes in regulation by prohibiting confiscation; and the Supremacy Clause provides additional protection by preventing the "trapping" of costs that might arise from conflicting federal and state regulations, when neither individually would be confiscatory in its effect.[8]

A contract between the government and a private party is referred to as a *public contract*. Traditionally, courts have been reluctant to find the existence of public contracts because they restrain future legislative actions and may thereby threaten the ability of governments to meet their sovereign responsibilities. However, recently in *United States v. Winstar,* 116 S.Ct. 2432, 1996 U.S. LEXIS 4266 (1996), the U.S. Supreme Court found that the Federal Home Loan Bank Board had established a federal contract with regulated entities—healthy thrift institutions—by inducing them to take over ailing thrifts in exchange for favorable accounting treatment of goodwill, and that, in passing the Financial Institutions Reform, Recovery, and Enforcement Act of 1989, Congress breached that contract by forbidding the favorable accounting treatment. Although the Contract Clause does not apply to the federal government, the Court held that the federal government should also be held accountable for breaching its contracts and pay damages to the acquiring thrift institutions because: (a) the federal government may not simply shift the costs of legislation that affects adversely its contractual partners by changing the law so as to prevent the bargained-for performance agreed to by the government, and (b) "from a practical standpoint, ... expanding the Government's opportunities for contractual abrogation, [would lead to] the certain result of undermining the Government's credibility at the bargaining table and increasing the cost of its engagements" (1996 U.S. LEXIS at 78). Courts should recognize similarly the existence of public contracts between state governments and regulated entities, such as local exchange companies, for which the Contract Clause provides remedies for governmental breach. With availability of a remedy for private parties, state governments will be compelled to consider carefully the sustainability implications of unilaterally imposed changes in legal and

[8]An example would be federal and state taxes on a firm's revenue that individually would not be excessive, but taken in combination would preclude its earning a reasonable return on past investments.

regulatory requirements; but at the same time, they will also be able to negotiate bilateral commitments that cover a broader range of performance obligations.

The results of the legal review depicted in Table 3.1 are reorganized in Table 3.4 to show the Constitutional limitations on unilateral and bilateral rules based on their transitionary effects arising from preexisting circumstances. The sustainability problems depicted in Table 3.4 all reflect the threat that changes in laws or policies will render the activities of private parties unremunerative after significant financial commitments have been made in reliance on previously governing rules and regulations. The preexisting circumstance most commonly addressed in the economics literature is the existence of a bilateral commitment in the form of a regulatory contract. Remedies for breach focus on compensation for expropriated investments. Goldberg (1979) stated that either government should be compelled to compensate for the diminished value of or inability to recover an investment, or a firm should be permitted to compensate the government to avoid the loss. In considering the threat to incumbent LECs' investments in backup capacity posed by entrants not similarly obligated, Weisman (1989) proposed two-part tariffs, whereby the option value to customers (and car-

TABLE 3.4

Constitutional Limits on Unilateral and Bilateral Rules Based on Preexisting Circumstances

Constitutional Clause	*Government Action Subject to Limitation*	*Preexisting Circumstances*	*Sustainability Concern*	*Remedy*
Takings & Due Process Clauses	Unilateral or bilateral rule.	Existing property investment; or investment based on existing bilateral commitment.	Sustainability of property rights system; or sustainability of existing bilateral commitment.	Invalidation of rule; or conversion of unilateral rule to a bilateral rule.
Supremacy & Commerce Clauses	Unilateral or bilateral rule.	Existing federal rule.	Sustainability of federal policy; or sustainability of firm.	Invalidation of rule.
Contract Clause	Unilateral or bilateral rule.	Investment based on existing bilateral rule.	Sustainability of property rights system; or sustainability of bilateral rule.	Invalidation of rule.

riers) of standby facilities is recovered on a flat-rate basis, sometimes referred to as a default capacity tariff (Weisman, 1988). Little guidance, however, is given as to how to manage the transition of an entire industry from one regulatory regime to another.

Yet, prior constitutional jurisprudence provides extensive experience in addressing expropriation problems under specific circumstances. The Takings and Due Process Clauses address expropriation problems as to existing real property rights and the sustainability of utilities under existing bilateral commitments. The Supremacy and Commerce Clauses address expropriation problems resulting from the conflict between federal and state rules, and the Contract Clause addresses those problems arising from conflicts with existing contracts, whether public (between the state and a private party) or private (between private parties). As such, the case law does provide us with some critical insights for addressing changes from traditional regulation to a more competitive environment in the United States and in other countries.

First, governments must recognize the existence of bilateral commitments with traditional telecommunications providers and anticipate the new confiscation problems that may arise from altering significant aspects of such bilateral commitments. The existing case law in the United States is based on confiscation problems that arose from ratemaking decisions. It is likely that new types of confiscation problems will arise with the elimination of the monopoly franchise. For example, in *GTE Northwest Inc. v. Public Utility Commission of Oregon*, 321 Ore. 458 (1995), the Oregon Supreme Court held that the state's Public Utility Commission's rules, which imposed physical collocation requirements without compensation on incumbent local exchange companies, constituted an impermissible taking of private property.[9] Other compensation problems may also arise from government attempts to asymmetrically impose cross-subsidy requirements and carrier of last resort obligations. Therefore, courts must be willing to grant remedies for these new types of confiscation by applying a more expansive reading of the principles underlying the Takings and Due Process Clauses.

Second, governments must be willing to renegotiate or establish new bilateral commitments as a whole. Piecemeal changes in regulatory rules may render existing, modified, or even new commitments unsustainable. This process can be facilitated in the United States if the courts are willing to in-

---

[9]The FCC's physical collocation rules were also stricken, but on the ground that the FCC exceeded its authority under the Communications Act of 1934, *Bell Atlantic Telephone Companies v. FCC*, 24 F.3d 1441 (D.C.Cir. 1994).The structure of section 251(c)(6) of the 1996 Act, which imposes physical—or at least virtual—collocation on incumbent local exchange companies, avoids a takings problem by providing such companies compensation from collocators pursuant to rates, terms, and conditions approved by the FCC.In a given situation, however, a taking may still occur if the FCC requires rates, terms, and conditions that are not compensatory.

terpret the Contract Clause so as to more readily recognize the existence of public contracts. This will expand the scope of bilateral commitments governments will be able to negotiate with private service providers while simultaneously forcing governments to address the sustainability implications of changes in laws and regulations.

Third, we should be more attentive to the ramifications of conflict between the rules of different governmental units. New types of interjurisdictional conflicts may arise with which we have little or no experience due to the rapidity of the transition from monopoly franchises to competition. However, this will likely require that the standards for determining the need for federal preemption under the Supremacy Clause in the United States (or, for example, the preemptive power of the relevant European Union directives in Europe) will be broadened. For example, the impossibility standard used in the United States for determining when federal laws and regulations preempt those of the states will need to be interpreted more broadly to include situations where, due to the complex interaction of different rules taken in combination, the "impossibility" of complying with both federal and state rules does not become apparent until a substantial period of time has passed.

SUMMARY AND CONCLUSION

New technologies and increasing competition are dismantling rapidly the system of monopoly provision of telecommunications services in the United States and in many other countries. Performance obligations once carried out as part of a bilateral commitment between service providers and governments are now being administered as unilateral rules. Yet, existing unilateral rules are fundamentally incompatible with a competitive telecommunications industry because: (a) they are applied differently to different firms; (b) firms have differential abilities and incentives to evade the rules due to difficulties in monitoring compliance; and/or (c) the investments required to satisfy the rules are sufficiently at risk to expropriation to preclude voluntary provision of service at desired levels of quality, continuity, and price. For long-term sustainability, the solution for the first problem is to apply unilateral rules symmetrically, for the second problem it is to convert unilateral rules to bilateral agreements, and for the third problem to convert unilateral rules to bilateral commitments.

To manage the transition to a more competitive regime, governments need to recognize the existence of bilateral commitments and be willing to renegotiate them as needed. Furthermore, the scope of governmental liability for breach should be broadened so that governments compensate firms for abrogating the terms of existing bilateral commitments. This liability should be based on a more expansive application of traditional, constitutional legal principles, relating to sustainability problems arising from preexisting investment or prior conduct, which underlie the Takings, Due Process, Supremacy, and Ex Post Facto Clauses of the U.S. Constitution. In this regard, governments need to anticipate new confiscation problems and new types of conflicts among the rules of different levels of government. In this way, policymakers will better address the financial effects of changes in regulatory rules on formerly monopoly providers and better position themselves to forge new bilateral commitments in the future.

REFERENCES

Bell Atlantic Telephone Companies v. FCC, 24 F. 3d 1441 (D.C. Cir. 1994).

Cherry, B., & Wildman, S. (1996, May). *A Framework for Managing Telecommunications Deregulation While Meeting Universal Service Goals.* Paper presented at the Second Annual Conference of the Consortium for Research on Telecommunications Policy, Evanston, IL.

Goldberg, V. (1976). Regulation and administered contracts. *Bell Journal of Economics, 7,* 426–448.

Goldberg, V. (1979), "Protecting the right to be served by public utilities," *Research in Law and Economics, 1,* 145–156.

Goldberg, V. (1980). Relational exchange. *American Behavioral Scientist, 23,* 337–352.

GTE Northwest Inc. v. Public Utility Commission of Oregon, 321 Ore. 458 (1995).

Haring, J. (1984, December). *Implications of asymmetric regulation for competition policy analysis.* FCC Office of Plans and Policy Working Paper No. 14.

Mueller, M. (1993). Universal service in telephone history: A reconstruction. *Telecommunications Policy, 17,* 352–369.

Noam, E. (1994). Beyond liberalization II: The impending doom of common carriage. *Telecommunications Policy, 18,* 435–452.

Panzar, J., & Wildman, S. (1993). *Competition in the local exchange: Appropriate policies to maintain universal service in rural areas.* Working Paper, Northwestern University, Evanston, IL.

Panzar, J., & Wildman, S. (1995). Network competition and the provision of universal service. *Industrial and Corporate Change, 4,* 711–719.

Shankerman, M. (1994, December). *Symmetric regulation for a competitive era.* Paper presented at the Twenty-Sixth Annual Conference of the Institute of Public Utilities in Williamsburg, VA.

United States v. Winstar, 116 S.Ct. 2432, 1996 U.S. LEXIS 4266 (1996).

Weinhaus, C., Makeef, S., Copeland, P., Calaway, G., Jamison, M., Hedemark, F., Harris, D., Monroe, T., Ralston, L., Bond, D., Inman, S., Albright, H., Dunbar, J., Garbanati, L., Sims, G., Adams, S., Mofils, J., & Little, L. (1993, July. *What is the price of universal service? Impact of deaveraging nationwide urban/rural rates.* Telecommunications Industries Analysis Project, University of Southern California, Boston, MA.

Weinhaus, C., Makeeff, S., Jamison, M., Albright, J., Garbanati, L., Calaway, G., Harris, D., O'Brien, M., Hedemark, F., Copeland, P., Sims, G., Adams, S., Sichter, J., Inman, S., Harrell, B., Thönes, R., & Connors, K. (1992, November). *Who pays whom? Cash flow for some support mechanisms and potential modeling of alternative telecommunications policies.* Alternative Methods Costing Project, Program on Information Resources Policy, Harvard University, Cambridge, MA.

Weisman, D. (1988). Default capacity tariffs: Smoothing the transitional regulatory asymmetries in the telecommunications market. *Yale Journal on Regulation, 5,* 149–178.

Weisman, D. (1989). Optimal re-contracting, market risk and the regulated firm in competitive transition. *Research in Law and Economics, 12,* 153–172.

Weisman, D. (1994). Asymmetrical regulation: Principles for emerging competition in local service markets. *Telecommunications Policy, 18,* 499–505.

Wiggins, S. (1991). The economics of the firm and contracts: A selective survey. *Journal of Institutional and Theoretical Economics, 147,* 603–661.

Williamson, O. (1976). Franchise bidding for natural monopolies—in general and with respect to CATV. *Bell Journal of Economics, 7,* 73–104.

4

Questions For Outlining A Universal Service Policy

Marlin Blizinsky
Institutional Network Project

This chapter lists questions that must be addressed in the universal service debate. Any universal service policy claiming to be comprehensive must answer these questions, and partial proposals must detail their relationship to the other listed questions.

I do not propose answers to these questions here. Rather, I encourage readers to keep these questions in mind as they read the other chapters in this volume, participate in the universal service debate, and evaluate the efforts of Congress and the Federal Communications Commission (FCC).

These questions have and will continue to be relevant even after Congress's recent passage of the Communications Act of 1996 (Sec. 254) and the FCC's adoption of new universal service rules (1997). Substantial work remains. The FCC's rules will likely be challenged in court and the rules leave substantial tasks to the states.

Additionally, these issues cannot be resolved permanently if the implementation of universal service depends on factors that change over time, factors such as the technologies that are in widespread use and our understanding of democratic participation (Blizinsky & Schement, chap. 5, this volume). In this view, the key universal service questions may be temporarily decided but not permanently resolved.

Finally, universal service policy is a set of subsidies. It costs money. Wrongly used, the policy becomes simply another entitlement (Murray, 1994), injures the increasingly important communications market, and harms us in other ways as well (e.g., Schrage, 1994).

RATIONALE

Should We Continue to Pursue a Policy of Universal Service, and, If So, Why?

At the core of the universal service debate is the issue of whether pursuing the goal of universal service is still desirable. The policy was first adopted in the Communications Act of 1934. Section 1 of the Act includes among its goals to "make available, as far as possible, to all the people of the United States a rapid, efficient, Nation-wide, and world-wide wire and radio communication service with adequate facilities at reasonable charges."

For many years the goal went unchallenged. That is true no longer. Respected observers have recently questioned the continuing validity of the goal. John Browning (1994) believed that universal service is an obsolete policy. He thought competition would soon drive down prices, ensuring that everyone who wants service will be able to afford it. Gilder (1992) reached the same conclusion but via a slightly different route. He thought bandwidth will be free because of technological advances.

Universal service should be retained only if there is a valid reason for doing so. The reasons may be political, social, or economic, a distinction that is not always clear. The reasons may be based on either theories of rights or the desirability of linking costs and benefits. And, these reasons may be based on factors accruing to individual recipients, network suppliers, or society at large.

Some commentators argue that certain goals, which they view as fundamental, can be accomplished only through access to communications technologies. They view this access as not only the ability to listen to the views of others but to do far more. For example, Emerson (1963) argued that "The right of all members of society to form their own beliefs and communicate them freely to others must be regarded as an essential principle of a democratically-organized society" (p. 883) The U.S. Supreme Court has taken a similar view, holding that "In the realm of, religious faith, and in that of political belief ... [the right] to persuade others to his own point of view ... in spite of the probability of excesses and abuses [are] essential to enlightened opinion and right conduct on the part of the citizens of a democracy" (*Cantwell v. Connecticut*, 1940, p. 310).

Others have focused on the perceived benefits from universal service and the perceived costs of lacking access. For example, it is far easier for a school to contact a truant's parent if the parent is reachable by telephone. It is harder to define who is benefited, and by how much.

Similarly, the benefits of access to the political process may accrue as much to society as to individual participants. The perceived inability to participate in the process may lead to social unrest, to defining laws resulting from that process as illegitimate, or to a decrease in compliance with those laws. If this is true, it could be argued that society and individuals both benefit.

The question arises, how can we give everyone the opportunity to form their own beliefs and to communicate them freely to others in a mass society? Some observers have concluded that for civil discourse to occur, citizens must have reasonable access to society's principal mechanisms for communication (House Report, 1984). Brown, Duguid, and Haviland

(1993) concluded that one of technology's roles is "to provide the opportunity and the resources for social debate and discussion" (p. 38).

Even if access to communications technologies is not necessary to meet key social or political goals, there may be economic benefits that make the policy worth pursuing. For example, Gilder (1994) reasoned that a network's functionality and value grows as the number of people connected to it grows. What is more, he believes this growth in value is exponential rather than arithmetic (Gilder, 1993).

Concluding that access to communication is beneficial does not, of itself, resolve the question of whether a universal service policy is desirable. As noted earlier, some observers believe the policy is unnecessary because of the decreasing per unit cost of memory and bandwidth. This view presupposes that the amount of communication capacity necessary to meet the identified goal or goals is constant over time. de Sola Pool (1990) and others have questioned this presumption. He has argued that the amount of bandwidth needed to be an active participant in society increases simultaneously as the cost per unit of bandwidth decreases.

Implementation of a universal service policy may cause telecommunications service market distortions so severe as to override the policy's benefits (Murray, 1994; Schrage, 1994). Much depends on the existing nature of the relevant market. If the relevant market is substantially monopolistic, the policy would, at worst, aggravate a bad situation. If the market is substantially competitive, serious misallocation could, but need not, occur. It is possible to design a policy that utilizes market allocations. The next chapter of this book presents one such design (Blizinsky & Schement, chap. 5). And, as discussed next, markets themselves may misallocate services when a substantial share of a service's benefits flow to individuals other than those paying for the service.

Finally, some values may be more important than market distortions. Few of us would suggest that airline safety should be market determined, for example. If the rights involved with universal service are fundamental, perhaps they are also too important to be left to the market.

Having balanced these factors, it should be possible to determine whether to retain the policy of universal service in some form or to abandon it. If one concludes that the policy should be retained, the factors involved in the determination may also inform questions about how such a policy should be implemented.

IMPLEMENTATION

Even if universal service as a policy is desirable, key decision must be made about the implementation of the policy. These decisions include: What service(s) should be included? How will the service(s) be made available? How should the service(s) be funded? and, What should government involvement be in decision making and implementation? These are difficult questions as Congress noted when it wrote, "One of the greatest challenges over the years in establishing communications policy has been assuring access to the electronic media by people other than the licensees or owners of those media" (*House Report*, 1984).

What Service(s) Should be Subsidized Under a Universal Service Policy?

When the Communications Act of 1934 was adopted, basic service meant telephone service (National Cable Television Association, 1993). There were no Internet, cable television, satellite, or wireless phone and data companies as there are today. Today it is less clear what we mean by universal services. Some observers continue to use the term to mean telephone service; others want the term broadened to include broadband and/or two-way high capacity services (Anderson et al., 1995).

Another approach is available—recipients could be given a dollar amount, in the form of credits or coupons, to spend among a list of alternatives (Blizinsky & Schement, chap. 5, this volume). One problem with this approach is that the choices recipients make may not be the ones society thinks best. Some commentators would be upset if people spent their communications coupons on cable television service, video rentals, or e-mail rather than on telephones.

The debate about what services should be included in the bundle is frequently sidetracked by participants' failure to distinguish between minimum basic service and equal service. Many people in our society lack basic communications services; too often, however, policymakers express concern about the need for creating equal access to modern technologies (see League of California Cities, 1994). These goals are not the same.

Equality is a much harder goal to achieve. In addition, attempting to reach this higher goal, we may fail to reach the lower. At this juncture, it is more important to define the minimum level of acceptable service, especially as we prepare the next generation of workers.

Where Should Services be Made Available?

Basic services can be made available in a variety of ways. For example, telephone service can be made more available by increasing the number and accessibility to pay telephones or by funding telephone services for every residence. Similarly, high-speed Internet access can be available at libraries, schools, public access centers, or residences over wire or wireless systems.

This decision is not necessarily simple. When Minneapolis attempted to increase access to telephones by increasing the number of pay telephones, it concluded that the telephones were being used for drug transactions as well as for ordering pizza. The city then modified certain phones so users could only place calls to police and fire services during late evening hours.

The issue of where services will be available is especially acute for people who live in rural, very high-cost connection areas. Should the subsidy be limited to the rural poor or given to all rural, high-cost households? It is simple to argue that Donald Trump's ski chalet should be excluded; it is less clear whether the residential development of the cook at the restaurant where Mr. Trump eats dinner after skiing should be excluded too.

How Should Universal Service(s) be Funded?

Whatever the service to be made universally available, someone must bear the costs. Any funding method should be fair and not distort the market.

The answer to the question of who should pay may depend on the answer to the earlier question of who is benefited by universal service. Those who are benefited should be expected to contribute. In such a scheme, government must pay for the share of the gain that goes to recipients and society at large.

The answer to the question of who should pay may be influenced by whether one believes there are major social, political, and economic externalities that accrue to society, network owners, or service providers from universal service. Cowen (1993) stated that "externalities occur when one person's actions affect another person's well-being and the relevant costs and benefits are not reflected in market prices" (p. 75). Externalities may be beneficial or harmful. According to Pearce (1992), "A beneficial eternality is where an eternality-generating activity raises the production or utility of the externally-affected party" (p. 146). As noted earlier, Gilder (1992, 1993) believed that there are extra benefits to network owners when users are added to a network.

Some economists think that when activities result in beneficial externalities consumers undervalue those activities because they will pay for the benefits they receive but not for the benefits received by others. Under this theory, some portion of universal service costs should be borne by all beneficiaries (Cowen, 1983).

A confounding issue is the debate over compensation for use of publicly owned rights of way. Support for compensation does not, of itself, mean that a provider should pay less toward funding universal service. If use of rights of way has value, there is no reason for lower payments to a universal service fund unless either the price for use is greater than the marginal cost of use or the compensation required is discriminatory.

Should Universal Service be Provided Directly to Recipients?

People may also differ about the form of universal service subsidies. Traditionally, recipients never handled the subsidy. The government paid the provider or it just went into the provider's cost and rate base.

That model works well in a situation where consumer choice is unavailable as it lessens transactional costs. It does not work as well in an era of competition or choice. In such a situation, some observers favor providing the subsidy directly to recipients (Blizinsky & Schement, chap. 5, this volume). The subsidy would be in the form of cash payments or vouchers similar to food stamps. A direct payment plan allows the recipient choice, utilizes the market to the extent there is one, and lessens, but does not remove, the need for the complex fact-finding proceedings required to determine the cost of service to providers.

Another market-based approach is to require service providers to bid for the right to provide service.

Some observers would argue that while a market-based plan is wonderful in theory, it is impractical. If universal service means local telephone service, only those with access to a local network are in a position to bid. Most residential customers are able to access service only from one provider (other than cellular providers that charge significantly more than landline companies). And, there is no agreement on whether such a plan would work if it utilized nonfacilities based competition. It may be that choice must wait for the development of such competition.

What Body(ies) Should be Entrusted With Making and Implementing the Policy?

Finally, we return to the role of government in decision making and policy implementation. Which level or levels of government should make the decisions? Should key elements of universal service be uniform among the states or is there room for experimentation? And, if there is to be experimentation, what is the federal role in decision making?

Should the same level of government make all policy decisions or can that responsibility be shared? Traditionally, key decisions have been made by federal and state governments. Local governments may be closer to citizens because they represent fewer people. If so, are they better positioned to make at least some of these decisions?

Even if the rules are uniform, which level(s) of government should be charged with implementing and enforcing those rules? And, is there a role for private industry in implementation? Could this function be contracted out?

Finally, do governments, as institutions, have interests that are separate from and, at times, even contradictory to, those of the public?

CONCLUSION

Universal service is not one decision, but many. A number of reasons can be brought forward to justify adopting or rejecting a universal service policy. But reaching the conclusion that such a policy is desirable does not end the debate—major implementation issues remain. These issues include:

1. What service(s) should be subsidized?
2. Where should service(s) be made available?
3. How should universal service(s) be funded?
4. Should universal service be provided directly to recipients?
5. What body(ies) should be entrusted with making and implementing the policy?

These questions are difficult, and honest people of good faith can disagree about their answers. What is clear is if we are going to advocate for and adopt a universal service policy, these questions must be addressed.

REFERENCES

Anderson, R. H., Bikson, T. K., Law, S. A., Mitchell, B. M., Kedzie, C., Keltner, B., Panis, & Pliskin, J. (1995). *Universal access to e-mail: Feasibility and societal implications.* Santa Monica, CA: Rand Corporation. Also available at: http://www.rand.org:80/publications/MR/MR650.

Brown, J. S., Duguid, P., & Haviland, S. (1993). Towards informed participation. In D. Bollier (Rapporteur), *The promise and perils of emerging information technologies, A report of the second annual roundtable on information technology* (pp. 25–39). Queenstown, MD: The Aspen Institute.

Browning, J. S. (1994). Universal service (An idea whose time is past). *Wired, 2*(9), 102–105, 152–154. Also available at: *http://www.wired.com/wired/2.09/features/universal.access.html.*

Cantwell v. Connecticut, (1940) *310* US 296 –311.

Cowen, T. (1993). Public goods and externalities. In D. R. Henderson (Ed.), *The fortune encyclopedia of economics* (pp. 74 –77). New York, NY: Warner Books.

de Sola Pool, I. (1990). *Technologies without boundaries* (1990). Cambridge, MA: Harvard University Press.

Emerson, T. I. (1963). Toward a general theory of the First Amendment. *Yale Law Journal, 72*, 877, 956.

Gilder, G. (1992). Into the fibersphere. *Forbes ASAP, 150*, 111, 125.

Gilder, G. (1993). Telecosm: Metcalf's law and legacy. *Forbes ASAP*, 152, 158–166.

Gilder, G. (1994). Telecosm: Life after television updated. *Forbes ASAP, 153*, 94, 105.

House Report on the Cable Communications Policy Act of 1984. House Report (Energy and Commerce Committee) No.98–934, Aug. 1, 1984, 5 US Code and Administrative News, 1984 Record Volume, 4655, 4667.

League of California Cities. (1994). *Telecommunications Policy.* Sacramento, CA: League of California Cities.

Murray, A. (1994, January 31). The outlook for Gore's data highway: Another entitlement?" *Wall Street Journal*, p. A1.

National Cable Television Association. (1993). *Cable television and America's telecommunications infrastructure.* Washington, DC: Author.

Pearce, D. W. (Ed.). (1992). *The MIT dictionary of modern economics* (4th ed.). Cambridge, MA: MIT Press.

Report & Order In the Matter of Federal-State Joint Board on Universal Service. (1997). FCC 97-157, CC Docket No. 96-45. Available at: *http://www.fcc.gov/ccb/universal_service/fcc97157/.*

Schrage, M. (1994, January 7). Let's not put the data highway on the list of our inalienable rights. *Washington Post*, p. G2.

III

SOCIETAL ROLES AND IMPLICATIONS OF UNIVERSAL SERVICE

5

Rethinking Universal Service: What's On the Menu

Marlin Blizinsky
Institutional Network Project
Jorge Reina Schement
Pennsylvania State University

Eletelephony

Once there was an elephant
Who tried to use the telephant—
No! No! I mean an elephone
Who tried to use the telephone.

—(Richards, 1955, p. 31)[1]

When the telephone was new, the very mention of its name could provoke laughter in children. Laura Elizabeth Richards' nonsense poem evokes the confusion of names that captivate children when they are first learning the intricacies of English. Yet as a metaphor, it also works on a level that provoked us on first reading it. Which way does the poem flow, from elephant to telephone or from elephone to telephant? Indeed, its very confusion is the essence of its charm. So too, confusion underscores the universal service debate. The basic goals and possibilities are unclear. Who should benefit? Why? Is universal service about the telephone?

---

[1]Laura Elizabeth Richards (1850–1943) was born in Boston. She was the daughter of Samuel Gridley Howe and Julia Ward Howe, author of the "The Battle Hymn of the Republic," which appeared in the February 1862 issue of the *Atlantic Monthly.* Laura was primarily known for her many books and poems aimed at juveniles, "Eletelephony" being one. She also authored *Five Mice in a Mouse-Trap* (1881), *Queen Hildegarde* (1889), *Captain January* (1890) (made into a movie in 1936 starring Shirley Temple), *Three Margarets* (1897), and *Star Bright* (1927). In 1915, she shared the Pulitzer Prize with her sister Maud Howe Elliott for a biography of their mother.

THE IMPORTANCE OF RECONSIDERING UNIVERSAL SERVICE AS A FOUNDATION POLICY FOR THE NATIONAL INFORMATION INFRASTRUCTURE

In the excitement over the national information infrastructure and the debate about increasing competition in the communications marketplace, some authors have questioned whether the goal of universal service adopted in the 1934 Communications Act[2] should be abandoned (e.g., Browning, 1994). Although Congress reaffirmed the goal of universal service in its passage of the Communications Act of 1996 (Sec. # 254),[3] it did little to clarify its meaning or application. Perhaps this was inevitable given the lack of consensus about the rationale for the policy, the proper components of such a policy, and the fairest way to fund it. Yet any meaningful discussion of universal service that is to lead to lasting implementation must include a rationale for the policy that makes sense to policymakers, as well as to the general public. That is the goal of this chapter.

Universal access to communications technology is a tool for enabling citizens to participate in the fundamental activities of society including its economic, political, and social life. The founders of our country understood that participation offered a multiplicity of benefits when, in 1789, the Congress mandated the first post road. In today's information society, it is crucial. More than ever, communications is the glue that binds us together.

Although the democratic principal is inclusion, the economic principle is contribution; to maximize the potential of each individual is also to maximize the nation's wealth absent evidence to the contrary. The costs and benefits of an inclusive universal service may be measured in the progress the nation achieves toward maximizing the contributions of each member and of the nation as a whole.

Lacking access to the community's key communications systems affects those people so situated. It makes it more difficult to do many things, ranging from getting the baby to the doctor to informing your employer that you are ill and will be absent. This situation enforces isolation and its derivative alienation.

Failure to support the contribution of each member negatively impacts the nation, too. The gross domestic product is lower. Potential wages are lost. The networks themselves are also less valuable. The more people on a network, the greater is the network's value to each member, to the network

---

[2]Section 1 of the Act states that the purposes of the regulatory structure are to "make available, as far as possible, to all the people of the United States a rapid, efficient, Nation-wide, and world-wide wire and radio communication service with adequate facilities at reasonable charges," to improve the nation's defense, and to improve the administration of communications regulation. P.L. 73–416, 73rd Cong., 2nd. Sess. Chap. 652, 48 Stat. 1064.

[3]Pub. L. 104–104, 110 Stat. 56.

owner, and to those providing services over the network. And, in the long run, increased government spending (and taxes) is needed to support unemployed or underemployed members of the community. Society benefits from having as many people as possible served by all of the available mainstream communication systems. Thus, for people without service, and for society, the stakes are high.

We set out to rethink universal service in response to persistent demographic conditions that marginalize too many in our nation. We also take note of changing market conditions and the convergence of key technologies. We have not addressed the cost, subsidy, or price structure of such a policy, although we recognize the necessity of doing so. Instead, we have chosen to lead with ideas. Our goal is to open up the policy discourse, increase the ferment of new ideas and, in so doing, point the way to greater possibilities for access and participation. In reality, this chapter reflects ideas in progress and it is in that spirit it was written.

THREE JUSTIFICATIONS FOR A UNIVERSAL SERVICE POLICY

Universal service provides three levels of value to U.S. society: political, economic, and social.

The Political Value

Democracy requires an informed and involved citizenry (Rotunda & Nowak, 1992).[4] This is possible only if its citizens have access to information about their government and the opportunity to participate in political discourse. First amendment scholars, such as Emerson (1963), have been particularly emphatic on this point. In Emerson's view, "The crucial point is not that freedom of expression is politically useful, but that it is indispensable to the operation of a democratic form of government. Once one accepts the premise of the Declaration of Independence—that governments derive 'their just power from the consent of the governed'—it follows that the governed must, in order to exercise their right to consent, have full freedom of expression both in forming individual judgments and in forming the common judgment" (p. 883). Not surprisingly, a succession of Constitutional amendments and Supreme Court decisions have sought to make the political process more open, or at least more open to a larger share of society.

Even if the importance of political information is obvious, it is extremely difficult to become an informed citizen. Watching television or reading the newspaper is not enough; the political process requires more than simply receiving news about politics and political issues. As the Supreme Court stated in *Cantwell v. Connecticut* (1940), "In the realm of religious faith, and

[4]See, for example, Vol. 4, Sec. 20.6, p. 15–16.

in that of political belief [the right] to persuade others to his own point of view in spite of the probability of excesses and abuses [is] essential to enlightened opinion and right conduct on the part of the citizens of a democracy" (p. 310). At the level of local democracy, political participation involves communicating with a public official, a fair housing advocate, or a representative of the American Association of Retired Persons. It requires citizens to register complaints with public officials, to inquire about polling places, and to learn the operating hours of public agencies.

There are, then, two dimensions of political participation—reception and distribution. On the reception side, citizens are better able to make informed contributions and to receive the benefits of the political process when they have heard a variety of opinions, especially when they have heard their favored opinions challenged in the marketplace of ideas. On the distribution side, citizens also benefit when individuals are able to communicate and to engage in political dialogue beyond the confines of their immediate communities. Only then, can democratic discourse transcend the walls of localness and the stifling of popular debate that occurs when only elites have access to the national channels of communication.

There are additional benefits as well. Democracies depend primarily on voluntary compliance rather than coercion to obtain adherence to laws and values. People are much more inclined to feel bound by and invested in the political process when they have effective involvement in that process. Conversely, alienation from that process is likely to lead to both anger and noncompliance (Emerson, 1993).[5] Free speech and participation thus provide a safety valve for society (Rotunda & Nowak, 1992).

So important are the many possible avenues for democratic participation, that the total of evolving information technologies and telecommunications policies must protect, facilitate, and expand democratic discourse.

The Economic Value

When the delegates to the constitutional convention met, they did so in part to resolve interstate economic conflict that hindered both interstate and international commerce. A major reason for the convention's actions was to change the form of the federal government to enable it to effectively address national commercial problems (Rotunda & Nowak, 1992).[6]

The document resulting from their deliberations did more than restructure the country's economic framework, however. Through the commerce,

---

[5]"The right of all members of society to form their own beliefs and communicate them freely to others must be regarded as an essential principle of a democratically-organized society. The growing pressure for democracy and equality reinforce the logical implications of the theory and demanded opportunity for all persons to share in making social decisions," p. 883.

[6]Vol. 1, Sec. 3.1, p. 299.

currency, post office, and other clauses, it granted the federal government authority to shape economic activities that positively affected individuals' as well as society's economic interests.[7] Thus, the constitution empowers the federal government to regulate bankruptcies, copyrights, and patents, and state limitations on mobility; requires the federal government to pay just compensation for private property taken; prohibits states from "impairing the obligation of contracts" including passing debtor relief acts and from refusing welfare benefits to new residents.[8] In addition, the first amendment protects commercial, as well as political, speech (*City of Cincinnati v. Discovery Networks, Inc.*, 1993).

A closer examination of these rights and limitations indicates that the Constitution envisions citizens as economic consumers and producers. For example, the Constitution protects the interests of consumers in quality and lower prices by preventing states from using discriminatory taxation to insulate local producers from competition (e.g., *West Lynn Creamery, Inc. v. Healy*, 1994).[9] The Supreme Court's decisions on commercial speech indicate a clear intent to protect the ability of producers to disseminate truthful information about their products (e.g., *Rubin v. Coors Brewing Co.*, 1995).[10]

In the information age, information networks distribute economic goods and services, and add value to transactions. Networks carry information that becomes input to other products or decisions, such as providing information on the current selling price of cattle to a rancher, and also carry information that itself is the product, such as participation in a commercial online discussion group.

Thus, the economic benefits of an interconnected information infrastructure accrue to the individuals on that network, to the network owners, and to society as a whole. For individuals, a telephone is needed to obtain most well-paying jobs and to call in sick; whereas, without one, it is difficult to order from a catalogue, have a pizza delivered, find the schedule for the next bus, and get homework assignments. Without basic communications service, a person is more likely to become an economic burden on society. A person is less likely to get a job when a potential employer cannot reach that person by telephone, and the person and their family suffer. The unem-

---

[7]E.g., U.S. Const. art. I, Sec. 8, cl. 3; cl. 5 and 6; cl. 7; also, art. I, Sec. 10 (prohibiting a state from imposing duties on imports, except for the purpose of funding inspections, without Congressional permission), and art. I, Sec.9, cl. 5. (taxes and duties on state exports).

[8]U.S. Const. art I, Sec. 8, cl. 4; art I, Sec. 8, cl. 8; Amendment 5; and art I, Sec. 10. See *Crandall v. Nevada*, 1867. On debtor relief acts, see Rotunda, and Nowak, p. 437. On refusing welfare benefits, see *Shapiro v. Thompson*, 1969.

[9]Invalidating a tax that applied to all milk dealers when the proceeds were given only to in-state dealers. Of course, this protects the rights of out of state producers as well.

[10]The Supreme Court has limited this right to products which are not deemed harmful.

ployed are also less likely to be taxpayers and more likely to receive social welfare payments.

When the number of people on a network grows, so does the network's functionality and potential customer base (Gilder, 1994).[11] A network is most valuable to owners, service suppliers and service recipients when as many people as possible connect to a network, all other factors being equal. Similarly, network users want as many services as possible and want the benefits of competition, such as innovation, lower prices, and improved quality and choice. Network owners and service providers especially benefit from universal service. Not only will a person without a phone have trouble getting a pizza delivered, the company selling delivered pizzas is likely to make fewer sales. As a new infrastructure develops, it will only reach its full potential when the maximum number of entrepreneurs is attracted by the participation on the network of the maximum number of people. Gilder (1993) believed that the value of adding an additional person to a network is not the added value brought by the person, but the square of that value.[12]

As we become an economy that increasingly creates and shares information, maximizing access to the interconnected information infrastructure carrying that information becomes crucial for businesses and individuals. To that end, universal service is the mechanism for maximizing access because Americans should be empowered economically by their telecommunications systems.

The Social Value

The writers of the Declaration of Independence included in their list of inalienable rights the pursuit of happiness, in the 18th-century meaning of personal liberty and fulfillment. In the 20th century, the Supreme Court has found that the right of liberty encompasses those privileges long recognized as essential to the orderly pursuit of happiness—and has further included the right to freely exercise religious beliefs, to marry, and to travel.[13] At the tip of the 21st century, it seems reasonable to argue that access to an interconnected information infrastructure is crucial because individuals need access to information for self-development, help in developing and

---

[11]Stating that, "computer cost-effectiveness rises by the square of the number of computers connected to the network," p. 95.

[12]Gilder has stated that Metcalfe's Law shows that when you "connect any number, 'n,' of machines—whether computers, phones or even cars—and you get 'n' squared potential value," p. 160.

[13]*Board of Regents v. Roth*, quoting *Meyer v. Nebraska*, (listing goals embodied in the Constitution's definition of liberty as including social goals) at p. 572; *Cantwell v. Connecticut; Boddie v. Connecticut;* and *Memorial Hospital v. Maricopa County.*

maintaining social relationships, and obtaining the benefits that come from those relationships.

Indeed, the range of information provided by the basic infrastructure is quite broad, encompassing the mundane and the critical, from the hours that a movie is playing and the location of the evening's party, to the call for a fire truck or police car. Contacting a relative to baby-sit, calling the library to check on a book's availability, and calling the toxic hotline all fall within the reasonable expectation of interconnectedness through the information infrastructure. Nor should the network be thought of as solely bound by the limits of telephony. Keep in mind that, in the social sphere, television is a key communications medium. Most people receive their news via television and social issues (big or small) are heavily debated in the medium. In fact, social chitchat often focuses on what is shown on television.

Access to communication services also offers benefits in a broader cultural sense. If the nation wants to encourage the sense of shared values and mutual responsibility that comes from social interaction, then maximum access to communication networks becomes a necessity. Social interaction forms a part of the socialization process through which society seeks to engender loyalty to itself and to prevent crime. We define ourselves not in isolation but through contact with others. Therefore, the network is an essential ingredient for overcoming social fragmentation and, consequently, for enabling participation in community. In essence, communication creates society.

Universal service, then, is an operational benchmark on the way to the greater goals of social empowerment, political participation, and economic development. Telecommunications policy should serve to connect all Americans to each other and to the rest of the world. Americans should experience benefits that can save lives, create jobs, and give every citizen the chance to pursue the full spectrum of life. In an information society, the network is the weave that helps us define ourselves and holds us together.

DEMOGRAPHICS AND UNIVERSAL SERVICE

Despite the importance of having access to basic communications services, approximately 6 % of the people in the United States still do not have access to a telephone (Federal Communications Commission, 1998).[14] And, there is considerable variation among the states in telephone penetration rates by household: from a low of 86.2% in New Mexico to a high of 97.5% in Connecticut (Federal and State Staff for the Federal-State Joint Board CC Docket No. 87-339, 1997).[15]

---

[14]Table 15.1. [/ 70. In comparison, 98.3% of households own televisions (U.S. Bureau of the Census, 1997, p. 566).

[15]Monitoring Report, May, 1997, Table 1.2, p. 21.

The variations are larger if comparisons are made between personal characteristics and telephone service. Reports provide the following view of households without telephone service.

Poverty

The lack of a telephone is strongly associated with income and race:

- Among adult heads of households between the ages of 15 and 24, 15% are nonsubscribers, the highest of any age group (Federal-State Staff, 1997);[16]
- Rates of nonsubscription for minorities of this age are even higher. 27.1% of households headed by people of Hispanic origin, and 22.7% of households headed by African Americans are without service. This compares with a nonsubscription rate of 13.8% for Whites within the same age range (Federal-State Staff, 1997);[17]
- Among households with annual incomes of less than $10,000, 14.6% are nonsubscribers and, as expected, this figure decreases with income increases. Again, aggregate figures mask the state to state variations. For this income group, nonsubscribership rates range from a low of 3.4% in Connecticut to a high of 31.5% in New Mexico (Federal-State Staff, 1997);[18]
- More than two thirds of those households without telephone service have annual incomes of $15,000 or less. (U.S. Bureau of the Census, 1994);
- Nonsubscribership among households headed by females with children living at or below the poverty line is approximately 50% (U.S. Bureau of the Census, 1994);
- Poverty, or low income, is a primary predictor of nonsubscribership. More than two thirds of those without telephone service have annual incomes of $15,000 or less. One of the noteworthy findings in recent analyses of census data on telephone subscribership is the very high rate of nonsubscribership among those households dependent on public assistance (U.S. Bureau of the Census, 1994);
- 17.6% of households in subsidized housing are without telephones (an increase of close to 2% from 10 years ago) (U.S. Bureau of the Census, 1994);
- 31% of households receiving food stamps have no telephone (U.S. Bureau of the Census, 1994);

---

[16]Monitoring Report, May, 1997, Table 1.7, p. 50.

[17]Monitoring Report, May, 1997, Table 1.7, p. 50.

[18]Monitoring Report, May, 1997, Table 1.9.

- 27.9% of households on welfare lack telephones (U.S. Bureau of the Census, 1994); and
- 43.5% of households that are completely dependent on public assistance lack a telephone (U.S. Bureau of the Census, 1994).

Mobility

The lack of a permanently owned residence also strongly correlates with the absence of telephone service:

- Renters are six times more likely than home owners to be without a telephone (U.S. Bureau of the Census, 1994);
- In New York State, renters make up 90% of the households without telephones (Department of Telecommunications & Energy, 1993);
- In those parts of California where subscribership fell below 90% more than half of the nonsubscribers had lived at their current address for less than one year (Field Research Corporation, 1993); and
- A person in-transit is less likely to have a telephone than a long-term resident (e.g., Chesapeake & Potomac Telephone Company, 1993; Rubin, 1993).

Privacy

Some people may prefer to not have a telephone because they believe it enables intrusions on their privacy:

- Some low income households may not subscribe to telephone service in order to avoid intrusion from unwanted sources. For example, Latinos in California report concerns about being reported to governmental agencies but these concerns rank well below other factors as reasons for not having phone service (Field Research Corporation, 1993, p. 90). Making telephone service more affordable may not bring these households onto the network; and
- Recent research has suggested that different social groups create varying combinations of media to meet their needs. For example, in some cases cable service is chosen over telephone service as a response to family circumstances (Horrigan & Rhodes, 1995; Mueller & Schement, 1995).

Disconnection

It would be inaccurate to say that most households without telephone service would prefer to be without service, however. The majority of those without telephone service once were subscribers (U.S. Bureau of the Census, 1994). Toll charges seem to be a key factor in nonsubscription:

- Of the nonsubscribers who previously had service, the principal reason for their current nonsubscription is their inability to pay toll charges, and this may be the single most frequent reason households are disconnected from the public switched network (Mueller & Schement, 1995);[19]
- Most customers involuntarily disconnected are above-average users of toll telephone service (U.S. Bureau of the Census, 1994); and
- Disconnection for nonpayment of toll charges is likely to occur disproportionately among low-income minorities (Chesapeake & Potomac Telephone Company, 1993; Field Research Corporation, 1993; Mueller & Schement, 1995).

There exists a strong positive correlation between income and subscribership. In addition, most households without telephones subscribed once and were disconnected from the network. Also, ethnicity seems to make a difference in subscription rates. And, finally, when forced by limited disposable income to choose among media, people make their selections based on their circumstances and the benefits they anticipate receiving.

For these reasons, the number of households at the edge of the information infrastructure is likely to remain large. Thus, any universal service policy must aim at including as many households as possible on the network, or risk the political, economic, and social cost of a large marginalized population.

COMPONENTS OF A NEW UNIVERSAL SERVICE

For any new policy of universal service to achieve its potential, the ideals of democracy in the information age must be balanced with the evolving diversity of the U.S. population and its changing information environments. The goals, components, outcomes, and values of the policy will have to synthesize those ideals and realities, if it is to maintain its relevance for the next 50 years.

Underlying Values

Information policies, principles, and regulations often rest on assumptions that go unnoticed or unexamined. Even in the majority of situations, policy directions are chosen within a set of beliefs that are contradictory, often unarticulated, and never fully accepted. The authors seek to avoid this to the extent possible.

---

[19]Interview data support the archival data gathered by the PacTel/GTE study.

Any universal service policy should aim to reconcile the following values while recognizing the impossibility of complete reconciliation (Redford, 1969):

Universality. The information infrastructure must offer interconnectedness as an opportunity to all Americans. The goal should be to reach all Americans within a reasonable length of time.

Affordability. Access to the information infrastructure must fall within the means of all Americans.

Interactivity. Interactive audio and video should be included in the basic technological standard.

Personal Choice. Americans should be allowed to personally select the configuration of access technologies and information services they thick best meets their individual needs.

Improvement. Any new broad social policy must lead the way to a better life.

Inclusiveness. The new universal service must be inclusive in order to maximize the benefit to the nation.

Continuity. The existing domestic electronic environment must be the technological foundation for further developments.

To be Reasonably Informed. Everyone should have access, at a reasonable price, to the information necessary for democratic, economic and social participation.

Goals

The following list of goals is based on the values articulated above, and should form the basis for a new policy. Funding goals are addressed elsewhere (see, e.g., chap. 4). Necessary goals include:

- Maximizing people's choice of services and channels included to enable individuals to create the information environment they think best meets their needs;
- Maximizing people's opportunity to access information for purposes of democratic participation, and the improvement of economic potential;
- Maximizing free and open communication among the largest number of people;
- Realizing the potential of the evolving information environments in which Americans increasingly live;
- Enabling people to take reasonable actions to maintain their network access;
- Responding to the demographic changes in U.S. society.

Program Components

Ironically, the intent of the 1934 Communications Act still applies up today: "To make available, so far as possible, to all of the people of the United States a rapid, efficient," communications network, "with adequate facilities at reasonable charges." To carry forth that intent requires recognition of the technological changes that have occurred since the Act's passage. Most importantly, new networks now integrate multiple media rather than having separate network each delivering a single service (e.g., Sprint Communications Corporation, L.P., 1998).

Reaching the new universal service policy goals requires translating those goals into actions designed to bring them about. The components of such a program must include:

- The addition of cable television, e-mail, and the Internet into the universal service bundle;
- The development of a menu of services reflecting the expanded bundle;
- The ability of eligible individuals to select from the menu, up to the limit of their subsidy, a combination of services that best meet their needs;
- The opportunity to change service selections within a reasonable amount of time;
- The identification of the individual as the basic unit of service, rather than the household; and
- The guarantee that an individual will have local access even if the individual is unable to pay his or her long distance charges.

In addition, the policy should pay special attention to levels of participation of different groups, in order to maximize the opportunity for all to participate.

Outcomes

Universal service programs should include a list of desired outcomes for measuring the program's success. The implementation of a program meeting the criteria described above should promote the following outcomes:

- Greater access to the total network, not just to local telephone service;
- Increased political participation;
- Enhanced labor force participation;
- Development of innovative network uses;
- Greater interconnectedness;
- Expanded development and adoption of new technologies;
- Increased learning, both formal and informal;
- Enhanced personal security;
- Improved public health; and
- Increased commercial activity and greater earnings.

Caveats

Not all of these recommendations are new. Some are already in place in a limited number of locations, but here the whole is greater than the sum of the parts. Still, there remain nagging questions that the policy community must continue to address. How do we protect privacy? How do we protect property?

Ultimately, universal service represents maximum connectivity in a policy whose goal is to create a unified nation in the dimension of cyberspace. In this sense, the National Information Infrastructure serves the same unifying purpose as did the roads of the Roman Empire, Britain's sea lanes, and the railroad tracks of the U.S. West.

HOW THE NEW UNIVERSAL SERVICE MAY CONTRIBUTE TO A NEW INFORMATION POLICY

Americans will increasingly use the media even as new categories of media are invented. The emerging picture is not one of convergence in the social landscape, however, but of divergence with immense variations in resources (National Telecommunications and Information Administration, 1998). A new universal service policy must respond effectively to this condition.

We should imagine the fulfillment of universal service as both micro and macro in its potential. At the micro level, a comprehensive universal service policy can enhance access and the quality of life for individual Americans no matter how poor or marginalized they might be. At the macro level, universal service offers a potent policy tool to advance democracy and the economic development of the entire nation. By keeping both of these goals in sight as we engage in the discourse on universal service, we will be in a better position to judge our collective responsibility as well as our individual opportunities.

Because of the growing importance of media, of the development of new media, and of the rise of multifunction networks, society must confront these complex and critical issues. Universal service should lead to freer and more open communications by the largest number of Americans. If we set ourselves the task of building and implementing a new universal service policy addressing the needs of all Americans, then social empowerment, political participation, and economic development will result. If, however, we decide these issues based solely on the short-term needs of the corporate and governmental players, then we forego the opportunity to build a more productive and equitable foundation for the information age.

In the past, Americans supported regulation of communications media as a method for bringing order and structure to the information environment of a particular technology. This was seen as advancing an undefined societal good. In this view, universal service represented simply the intent to wire the nation. To think otherwise was unimaginable.

The authors now suggest that the welfare of society may be better achieved if people actively shape the content of universal service for themselves. Such a proposal also borders on the unimaginable just as wiring the nation did in 1934. Yet, as we increasingly balance the technological opportunities available to us against the burdens of shifting demographics and poverty, we must stretch our imaginations.

The telephone no longer evokes children's wonder; that aura has passed on to other technologies. Indeed, the aura itself has changed, for, today, we rarely if ever write poetry about any information technologies. Yet the vision of a connected people remains with us. Thus, this chapter represents a work in progress.

To reimagine universal service is a daunting challenge and, so, we have not attempted to define a definitive solution here. Instead, this chapter should be taken as an invitation to discuss and to imagine. To imagine a new universal service is to ask, not where we have been nor who we are, but who we want to be. Our imagination must not be limited by the constraints of today's technology nor by the existing orbits of the nation's corporations; for we are in short supply of the means by which to enable and to achieve the value and benefits of a fully participatory democratic society. In its best tradition, universal service is a discussion; let us continue it.

ACKNOWLEDGMENTS

This paper was presented in a slightly different form at the Twenty-Fourth Annual Telecommunications Policy Research Conference, Solomons, Maryland, October 5–7, 1996.

The authors wish to acknowledge the invaluable contributions of Larry Povich of the Industry Analysis Division of the Common Carrier Bureau of the FCC, and Milton Mueller of the Department of Communication in the School of Communication, Information and Library Studies at Rutgers University.

REFERENCES

Board of Regents v. Roth. (1972). 408 U.S. 564, 33 L. Ed.2d 548, 92 S.Ct. 2701.

Boddie v. Connecticut. (1971). 401 U.S. 371, 28 L. Ed. 2d 113, 91 S.Ct. 780.

Browning, J. (1994, September). Universal service (an idea whose time is past). *Wired*, 2.09, 102. Also available at: *http://www.wired.com/wired/2.09/features/universal.access.html.*

Bureau of the Census (1994, November). *Current population survey.* Washington DC: U.S. Department of Commerce.

Cantwell v. Connecticut. (1940). 310 US 296, 310.

Chesapeake & Potomac Telephone Company (1993, October 1). *Submission of telephone penetration studies in Formal Case No. 850.* Washington DC: Washington, D.C. Public Service Commission.

City of Cincinnati v. Discovery Networks, Inc. (1993). 507 U.S. 410, 123 L. Ed. 2d 99, 113 S.Ct 1505.

Communications Act. (1934). P.L. 73-416, 73rd Cong., 2nd Sess. Chapt. 652, 48 Stat. 1064.
Communications Act. (1996). Pub. L. 104–104, 110 Stat. 56.
Crandall v. Nevada. (187=67). 73 U.S. (Wall) 35, 18 L. Ed. 745 (1867).
Department of Telecommunications and Energy, City of New York (1993, November). *New York City household telephone penetration study, A report on the status of universal telephone service in New York City's neighborhoods*. New York: Department of Telecommunications and Energy, City of New York.
Emerson, T. I. (1963, April). Toward a general theory of the First Amendment *Yale Law Journal, 72*, No. 5, 877–883.
Federal Communications Commission (1998, July). *Trends in telephone service*. Washington, DC: Federal Communications Commission.
Federal–State Joint Board Staff, (1995, May). *Monitoring report*. CC Docket. No. 80–286.
Federal–State Joint Board Staff, (1997, May). *Monitoring report*. CC Docket. No. 87–339. Also available at: *http://www.fcc.gov/Bureaus/Common_Carrier/Reports/FCC-State_Link/Monitor/mr97-0.pdf*
Field Research Corporation. (1993). *Affordability of telephone service*. Washington, DC: Pacific Telesis.
Gilder, G. (1993, September 13) Telecosm: Metcalfe's Law and Legacy. *Forbes ASAP, 152*, No. 6, 158–166.
Gilder, G. (1994, February 28) Telecosm: Life after television updated. *Forbes ASAP*, Vol. 153, No. 5, pp. 94–105.
Horrigan, J. B., & Rhodes, L. (1995, September) *The evolution of universal service in Texas*. LBJ School of Public Affairs, University of Texas at Austin, Austin, Texas.
Memorial Hospital v. Maricopa County. (1974). 415 U.S. 250, 39 L. Ed. 2d 306, 94 S.Ct. 1076.
Meyer v. Nebraska. (1923). 262 U.S. 390, 67 L. Ed 1042 43 S.Ct. 625.
Mueller, M. L., & Schement, J. R. (1995). Universal service from the bottom up: A profile of telecommunications access in Camden, New Jersey. *The Information Society, 12*, 273–292. Also available at: *http://ba.com/reports/rutgers/ba-title.html*.
National Telecommunications and Information Administration (1998, July). *Falling through the net II: New data on the digital divide*. Washington, DC: U.S. Department of Commerce.
Redford, E. (1969). *The regulatory process*. Austin, Texas: University of Texas. Press.
Richards, L. E. (1955). *Tirra lirra*. Boston: Little, Brown and Company.
Rotunda, R. D., & Nowak, J. E. (1992). *Treatise on constitutional law: Substance and procedure*, (2nd ed.), St. Paul, MN: West Publishing Company.
Rubin, S. J. (1993). *Telephone penetration rates for renters in Pennsylvania*. Harrisburg, PA: Pennsylvania Office of the Consumer Advocate.
Rubin v. Coors Brewing Co.. (1995). 517 U.S., 131 L.Ed.2d 532, 115 S.Ct. 1585.
Shapiro v. Thompson. (1969). 394 U.S. 618, 22 L.Ed.2d 600, 89 S.Ct. 1322.
Sprint Communications Company, L.P. (1998, June 2). *Sprint unveils revolutionary network*, Press Release (June 2, 1998). Also available at: *http://www.sprint.com/sprint/press/releases/9806/9806020584.html*.
U.S. Bureau of the Census (1994). *Current population survey*, Washington, DC: Author.
U.S. Bureau of the Census (1997). *The statistical abstract of the United States* (117th Edition), Washington, DC: Author.
West Lynn Creamery, Inc. v Healy. (1994). 512 U.S. 186, 129 L. Ed. 2d. 157, 114 S.Ct.2205.

6

The Social Architecture of Community Computing

Allen W. Batteau
Wayne State University

PROLOGUE

A spectre is haunting North America. It is the spectre of ubiquitous information—anything, anytime, anywhere. The foundations of existing institutions—government, corporate, academic—crumble as legions of networkers use the power of information to warp time and space, interpersonal context, and social verities. New institutions and new leadership in the society will emerge only among those who understand where this information revolution is carrying us.

BACKGROUND: THE PROBLEM

This chapter begins in a melodramatic fashion, with an allusion to Marx, to highlight what I see as a critical problem in our discussion: the nature of the information "revolution" that is sweeping our country. It is undeniable that vast changes are taking place in digital technology and in the corporate, government, and academic institutions that are its primary users and sponsors; one need only look at the restructuring of the entertainment/telecommunications/computer industry to appreciate this. Whether this is revolutionary, or merely a continuation of trends that are nearly 100 years old, is a subject of perhaps ultimately academic debate. Whether this revolution is all hype, as *Wired* magazine tells us, or a set of developments that will change the way Americans live, work, learn, and consume, as Bill Gates tells us, dictates the character of one's involvement, sitting on the sidelines or engaged in the development of cyberspace. Whether this will create new possibilities for democracy, and whether or not that is something we really want, is a dilemma that should concern us all.

In an effort to answer some of these questions we have launched two experiments in trying to bring computing into inner-city neighborhoods in Chicago and Detroit. We have launched these experiments for several reasons: the inner city presents the greatest social distance between technology developers (suburban, middle-class, university educated) and technology users (poor, minority, uneducated), and as such presents an incredible laboratory for testing hypotheses of the relationships between technology and society. Second, the potential of this technology for dividing us into a nation of information haves and have-nots is an issue that should be of concern to all Americans. And finally, we saw in the inner city not poverty and pathology, but an underserved market. We believed that, if the right information resources could be created, the inner city would finance its own universal access. I want to stress that last point because we have launched all of these initiatives as business ventures, enterprises that would be judged on their ability to generate a profit.

I should acknowledge here that our work owes a great debt to my colleague Jorge Reina Schement. Dr. Schement's studies of telephone service in the inner city neighborhood (Schement, 1995; Schement & Mueller, 1995) and how residents creatively use telephone service to improve their lives, has been a constant guide for us.

Following in Dr. Schement's footsteps, we found that there was a market for advanced information technology and resources in the inner city, although it does not mirror the middle class or suburban market that most of us are familiar with. We, too, are still attempting to develop an adequate characterization of the market.

To expand on that point as a fundamental statement of my problem, we lack an understanding of the demand characteristics and potential uses for advanced information technology and resources in the inner city. We know that families there have substantial information budgets, that some do purchase computers, and that their interests in content go well beyond entertainment. Beyond that, it gets quite murky, with but a few points of light in the form of locally specific anecdotes to illumine the landscape.

COMMUNITY COMPUTING TODAY

There are a number of initiatives around the country that fall under the rubric of "community computing." I organize this around computing because in our society the computer, in both its technological and totemic statuses, appears to be the driver of information usage in our society today. The next 10 years will see an even greater convergence of computing and telecommunications technologies, and of the use of this new technological hybrid to bring numerous applications and forms of content into the home, school, and workplace.

One form of community computing is the Freenet movement, which has a stronghold in the midwest and in California. Freenets are the digital equivalent of public access channels on cable TV, which is to say they are not free, but rather piggybacked on top of other storage and processing, where the opportunity costs are practically free. Freenets are essentially Bulletin Board Services (BBS) and electronic mail (e-mail) systems, in which anybody in a given territory can get a subscription, post messages (including commercial messages), or create a discussion group. Many—perhaps most—Freenets are university- or library-established, and are used to publicize their services. Few that I have been able to discover have full Point-to-Point Protocol/Serial Line Internet Protocol (PPP/SLIP) connectivity, and hence cannot offer graphical access to the World Wide Web (WWW).

(This, incidentally, illustrates an interesting and problematic point. Freenets began to take off just as Mosaic and the Web became viable. Graphical user interfaces (GUI) offer a broader range of authoring capabilities, and I would suggest make computer-mediated communication far more visually accessible than ASCII text displayed in hardware fonts. Even elements so simple as proportional fonts, bars, or shadow boxes heighten the visual appeal of a message. Freenets that do not upgrade to this technology will become silent movies in the age of the talkies.)

Use of a Freenet, however, requires one to own a computer, or else travel to a library or other institution that makes computers available to the public. One of the earliest Freenets, the Santa Monica Public Access Network, established kiosks and even allowed the homeless of Santa Monica to have e-mail addresses. Before the "netbozos," angry individuals whose rantings dominated the bandwidth of the network, took over, this had some great benefits for Santa Monica as a community: It improved communication between the homeless and other residents of the city, it allowed those who had been previously voiceless to articulate their needs, and it led to the creation of some innovative city services for the homeless. The ultimate fate of Santa Monica Public Education Network (PEN), where a few angry voices (who were attracted to the new medium as an opportunity to "express themselves") crowded everyone else off the air, is also illustrative of the limitations of this technology: It created new public spaces, but like all public spaces, in the absence of policing (a system administrator or moderator), Gresham's law took over. Bad tokens drove out good.

A second form of community computing is the development of institutional networks at the community level. In every major city, schools, hospitals, social service agencies, and churches are setting up networks and creating computer facilities. The Detroit public schools are creating a system-wide network, with ISDN connectivity to every school and 56 KB lines from the schools to a central facility. Anticipating a major fall-off in traffic after 3 p.m., they are looking for ways to open up this to other institutions. This story is being repeated all over the country.

A limitation of this approach, offered here not so much as reproach but as caution, is a too-great reliance on hardware push: an "if we build it they will come" sort of view. Well, some will come, but others will not. We know from extensive experience in business that, if the potential of computing is to be realized, an adequate training and support environment must be provided. The costs of training and support actually exceed the costs of end-user devices. When this point is not understood, one finds rooms full of hardware sitting around unused.

One other activity that is worth mentioning is the cybercafes that are springing up around the country. The one that I am familiar with, in Birmingham, Michigan, offers Internet access at $10/hour, plus capuccino at $5 per cup. They are still developing a viable business model, and right now are supporting themselves by selling Web home pages at $2,000 each.

Also worth mentioning, because our subject is "community," is the idea of a virtual community. These ongoing online conversations have been discussed at length elsewhere; they provide social participation to numerous individuals, and forms of interaction that are especially valuable to dispersed or isolated individuals with shared interests. Whether or not virtual communities can be extended much beyond their current domains, or whether there is any value to this, are interesting questions.

Some conclusions can be drawn from the current level of activity in community computing. First of all, there is strong demand and interest, although sometimes this demand is more directed at the computer-as-totem rather than the computer-as-technology. One can expect to find continued strong growth.

Second, hardware costs are only a minor issue. Given the declining unit costs of storage and processing power, we are very close to the point where it would be to the telephone companies' advantage to give away low-end computers with 2400 baud modems. Whether or not the regulators would let them do this, and how these devices would be used, is another matter.

Third, more important than hardware costs is a much larger and modestly intimidating set of issues, whose trends are not as auspicious as those of hardware costs. These include:

1. processor and memory requirements of newer applications,
2. GUI and performance expectations of users,
3. training and support requirements for advanced hardware and applications, and
4. competition with other content in the users' lives.

The success of community computing will depend on the ability of user groups to finesse these and similar issues.

UNDERSTANDING THE COMMUNITY

Community is perhaps the most misused term in popular sociological discourse today. When an activist talks about "the community," that person is putatively referring to something genuine, out in the neighborhoods, in contrast to the downtown artificiality of government and corporate institutions. When a civic leader talks about "our community," he or she is painting a picture of inclusiveness. Other usages include a reference to occupational, technological, or interest commonalities, such as the scientific community, the UNIX community, or the First Amendment community.

The sorts of communities I am interested in are communities in place; that is, groups of people whose ties are based, among other things, on geographic propinquity, making possible multistranded, face-to-face affiliations. Usually this is less confusing to people who live in a community than to the sociologists who study their lives. Having established this minimal definition, there are multiple perspectives on the community, several of which are critical for establishing the community context of universal service.

The first perspective is community demographics. Although one sometimes associates a community with some sort of demographic similarity, there are many exceptions to this. In some suburban tracts where zoning codes and homeowners covenants are strictly enforced, communities represent a very narrow demographic slice; absent such mechanisms, heterogeneity is more typically the norm, either of race, ethnicity, occupation, or education.

Nevertheless, communities do have demographic profiles, which establish a set of constraints for any place-based computing strategy. Low-income communities will be more cautious in their hardware purchases; Hispanic communities will require Spanish-language services; less-educated communities with low mean educational attainment will require different forms of support and documentation from more educated communities.

One community that we conducted a study in was the South Chicago neighborhood. Comprising 3.3 square miles and having a population of approximately 41,000, this community was 62% Black, 32% Hispanic, and 7% White. This community has an unemployment rate of approximately 20%, due primarily to the shutdown of the South Works, a steel plant adjacent to the neighborhood. Approximately 40% of the residents were identified by the census as below the poverty level.

These demographics affected the community's relationship with computing, but in some unexpected ways. The effect might be summarized as the cultural magnification of demography. In this predominantly minority community, for many computing was seen as a "White" thing; in this poorly educated community, many told us that they did not think they could use the computer because they were not well educated.

While discussing the demographics, I should also put to rest one issue. This was a low-income community. Yet in our survey we found that the typical family spent $170 per month on information, including print and broadcast, telephone, and other communication. Families also said they were willing to spend more, an average of $16 per month, if they thought it would benefit their children's education. These numbers make it clear that, if networked computer-mediated communications were to replace other parts of their current consumption of information, then the costs of the basic infrastructure—transport and end-user devices—are within their reach.

Although the community demographic profile establishes a set of boundary conditions, it is clear that they are not as stringent as one might imagine. A second characterization of the community is in terms of its institutions—the churches, schools, civic associations, and block clubs that form an important part of people's lives. Indeed, these institutions provide a far better understanding of the community than do geography and demography. First of all, people do not necessarily associate with their neighbors, particularly if there are racial or status differences between them; they will associate across status lines within the churches. (The churches, incidentally, established a fundamental structure of the community; Blacks were Protestant, Hispanics were Catholic.) Elementary schools were a truly community institution, broadly embracing the neighborhood yet also incorporating some of its basic divisions.

These institutions were interested in our work. They had a more educated leadership, many of whom had at least a passing awareness of current developments in information technology; they were interested in seeing how this new technology could aid their institution. Although more than 80% of the respondents to our survey (conducted in June 1995) said they had never heard of the information superhighway, all of the institutional leaders we worked with had.

A third perspective on the community comes from an identification of common needs. Here the data and the voices of the residents were unambiguous. The big four needs were education, jobs, safety, and housing. Various institutions and community associations had programs in place to address these issues, although their resources were never adequate. It would be foolish to think that technology would solve this problem, particularly in a competitive society. It is equally foolish to think that information and technology are irrelevant to the solution.

From an information perspective the needs that community residents identified were:

1. What jobs can I do?
2. What work is out there?
3. How can I get it?

For the typical South Chicago resident, these questions were genuine puzzles; the limitations on their understanding of the world were limitations of both education and culture.

A final perspective to introduce is that of culture or subculture. Communities have subcultures; if they did not, they would not be communities. A community's culture provides the lenses through which they perceive the world: is a misfortune a crisis or an opportunity? Are their neighbors essentially deserving of their trust, or otherwise? A community's culture holds the community together; it is the safety net, better than any government program. It provides a floor through which none may fall, although it also creates a ceiling above which members should not rise. The community's culture is a source of stability and cohesion within a resource-poor and chaotic environment.

One cannot mention culture without introducing two other terms: power and investment. Community cultures are related in complex ways to the degree to which members of the community have, or feel they have, control over their own lives. Likewise, a community culture is a consequence of the investment of lives and livelihoods in the community's institutions. When families spend their evenings helping fellow parishioners, when parents volunteer at their children's school, when neighbors work together to clean up their street, they are investing in their community and developing its culture. They are willing to do so because they anticipate some return from their investment, if not in their lives then in their children's. On the other hand, if they do not see that such investment will yield any return, or if they expect more powerful forces to take it away from them, then they will not make the investment, and the community's culture will be, at best, weak and defeatist.

EMPOWERMENT WITH INFORMATION

An alternative vision would be to create uses of information that are empowering. These could range from educational applications to personal and community resources, such as information on job availability and neighborhood development strategies. In our research we found that education and housing were high on the list of information wants that a new information resource could supply. With a modest amount of cleverness, there could be a profitable business in providing information empowerment.

The cost of information empowerment is a nonsensical question. Empowerment can neither be sold nor granted; it can be recognized, and perhaps cultivated. None of us can go into an inner-city community and provide information empowerment, but we can provide the tools.

We learned several things in our efforts to bring these tools to some Chicago neighborhoods. The first lesson is the critical issues for information access, summarized in Fig. 6.1. The six issues—content, transport, network

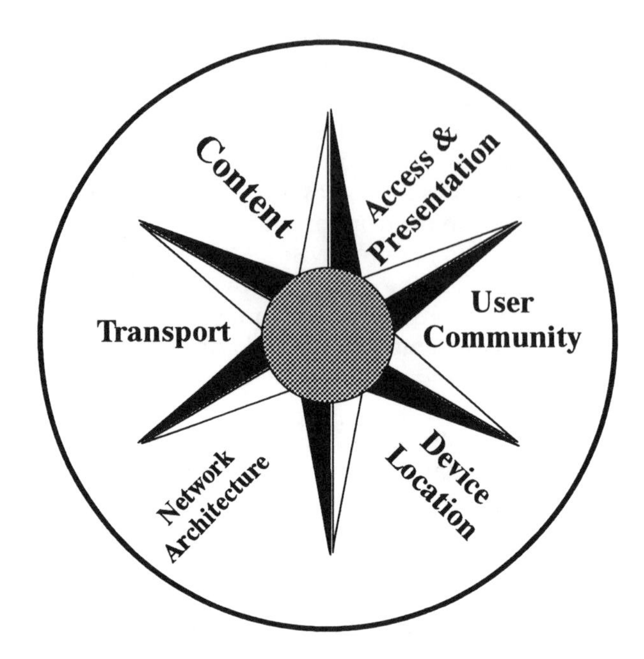

FIG. 6.1. The compass rose of the information landscape.

architecture, device location, user community, and access and presentation—are the interconnected issues in a sociotechnical architecture of advanced information services.

The first of these, *content,* originally seemed the most important. Content refers to what sort of information can be acquired through the system, in what format. We found that graphical content was more attractive than text-only, although, if the textual content had an emotional context (such as e-mail from a close friend), it was also attractive.

Transport is an issue on the technical side of the sociotechnical system. It refers to the physical infrastructure for the delivery of content: whether twisted-pair, coaxial cable, or fiber-optic cable. Although the technology (of modems and data compression) associated with this is changing rapidly, the demands placed on transport (by files including bitmapped images, animation, and sound effects) are also escalating rapidly. In this we see one more example of how the performance benefits of a new technology (such as high-speed modems) are frequently offset by escalating social demands on the technology.

Network architecture is the configuration of servers, routers, and transport that connects the end users to each other. This is another technical issue with a social dimension: There is no demonstrable requirement for all users to have identical access to a network backbone. For many applications smaller amounts of bandwidth are perfectly acceptable. The information superhighway is more a collection of roads, some four-lane, some arterial, and some residential streets with lower speed limits.

Network architecture, however, places severe constraints on the *location of end-user devices*. As long as the switched network was used only for voice messages, this was a minor concern: end-user devices were Private Branch Exchange (PBX) and telephone handsets in homes and businesses. With new forms of content, however, comes a proliferation (and stratification) of end-user devices: Internet computers (a computing device lacking a central processing unit, essentially a web-enabled terminal), business transactions, and institutional applications (such as digital libraries), all require potentially higher levels of bandwidth than the ubiquitous residential line. This issue is resolved by locating the higher end uses in common locations such as public libraries and educational institutions.

The location of these devices is closely related to the *user community*. Although much of the imagery of information use focuses on individual uses, in *every* case the use of information defines a community: Whether voice conversation within a family, business transactions within an industry, or the pooling of information within a local group, in all cases the production and use of information is not a solitary act, but rather the creation and constitution of a social relationship.

Communities have and create cultures. The 70-year history of the transformation of U.S. culture through the use of electronic media now finds its latest chapter written on the Internet, with the growth of a network culture and the new media of the web. A community's culture provides one of the major filters for the production and perception of information. Culture creates a screen for the *access and presentation* of information. Certain messages are filtered out because they are inconsistent with preexisting cultural norms; other messages that reinforce cultural assumptions gain instant credibility and travel across the social network rapidly. From a cultural perspective (as contrasted to a technical perspective), the community has the most access to information that is least likely to provoke a cultural change. (Presentation is the other side of access: how content is actually presented often affects its credibility.)

A second lesson concerns the presence and criticality of local information leaders. In a corporate setting, we characterize these people as champions or change agents. In the community, these people are the champions for change: individuals who have bought a computer for themselves, and are now eager to show everyone else what they can do with it. We found that, without information leaders, there was very little interest among the residents in experimenting with the new technology.

Finally, there are lessons concerning the external context. How information services are brought to urban communities is going to be partly driven by forces outside those communities. I am convinced that universal service within an urban community is a profitable business opportunity that does not require any sort of public subsidy or public mandate. We found that South Chicago residents were aware of, had access to, and paid for goods and services such as indoor playgrounds, fashionable clothing, cellular telephones, home entertainment centers, and fast-food restaurants, which are usually seen as highly discretionary. Some of the business models for a community-based information resource would include Kinko's, which provides a business service from a neighborhood location, and Kids Playworld, a chain of indoor playgrounds at which parents pay $5 per child. Deploying such service does require a willingness by technology developers to experiment with applications and information resources tailored to inner-city needs. If federal agencies can develop "tools for suburban schools" (in the words of one program officer), the least they can do is provide similar support for the inner city as well.

REENGINEERING THE INNER CITY

The recognition of a broad spectrum of municipal and inner-city resources offers the entry point for developing a community-based information strategy and preparing a community for the full benefit of the information superhighway. The first step is hence to identify and evaluate the resources that are available in the community and in the city. In the community these resources can include local businesses, churches, schools, community development organizations, political clubs, block clubs, and neighborhood associations, as well as less formal aggregations. In some communities, for example, the post office represents an informal gathering place that is used for information exchanges. All of these resources have their own agenda, and all will vary in their capability. Related to this is building a small number of central partnerships. These will be with organizations that are willing to invest their time in creating the information strategy. Community development organizations are a logical choice here, although, in some communities, schools or churches may be equally appropriate. The critical element is that the leadership of the organization has the sophistication to understand some of the implications of the information revolution. Part of building these partnerships is to establish and articulate the common goals of the partnership. A content company, for example, might have a common interest with community service organizations in selling information services that assist low-income families in finding housing. In this process, other critical partners will be found, such as municipal housing agencies.

These partners can cooperate in the assessment of the current information environment. Here it is essential to know what information resources the community has (and the credibility given to the different resources), what additional information resources are currently thought to be useful, and what information-intensive tasks the community faces. This information profile provides the baseline for development: if an individual's reading matter begins and ends with the *National Enquirer*, then a free subscription to *The Economist* will probably not enhance his understanding of world affairs.

As information-intensive tasks are identified, they need to be prioritized. These priorities must belong to the community. The strategy here is not to solve every problem, but to supply thoroughly the critical information for solving the most pressing problems. Our interview information suggested that job seeking will be very high on this list; this identifies what types of information (content) need to be made available.

Once we understand what sort of information is critical to the community, we can use knowledge of both here-and-now and future technology to supply it. For example, one could imagine a set of integrated services, available at devices placed in central community locations:

1. Assessment of personal skills and aptitudes.
2. One of any number of interactive assessment tools, perhaps reconfigured as an animated video game. The user would respond to a set of questions/challenges, and the system would provide a profile of skills, aptitudes, and motivations. The data could be ported over to the next application.
3. Information on job categories that match personal skills.
4. A lookup table from the *Dictionary of Occupational Titles* matching the skill/aptitude profile with job titles; or alternatively matching employment goals with skill requirements. The data could be ported over to the next application.
5. Information on skill improvement opportunities.
6. A database indexed by location, job title, and other criteria. Using a touch screen, the user could draw a circle on a map of the city, and the system would return a list of all training opportunities matched to the user's employment goals and within his or her commuting radius.

These low-end applications have nothing to do with the information superhighway. They can be put in place today, using the existing communications infrastructure and 5-year-old computing technology. As these are planned, more futuristic enhancements can be visualized as well. Remaining within our domain of employment information, one can suggest some additional information that could be delivered over the next generation of communications services:

1. Three-minute motivational video clips of different job opportunities.
2. Download of current, public domain data from state employment and tax agencies, showing which employers are adding payroll and which are shedding employees.
3. An employment-oriented multi-user dungeon (MUD) integrating in real life (IRL) employment experiences and accomplishments with scoring points in a video game.

These, of course, are flights of fancy, whose benefit to the community may be extreme, or nil. Only the community can decide. The general points are: (a) that the strategic information resources available to the community can be greatly enhanced using existing capabilities, and (b) that the community's strategy should consist of both exploiting current capabilities and planning for future capabilities.

There is an important agenda here: Only by exploiting existing technological capabilities for strategic purposes can the community develop the skills and sophistication to make strategic use of future capabilities. Information mastery (not just computer literacy) must be one more strategy in the toolkit of community development. In the absence of cultivating and growing the information skills and sophistication of the community, the information revolution will bring nothing more than a wider variety of entertainment and 500-channel home shopping.

Finally, such strategic planning of community information resources should address multiple-task domains. Having several irons in the fire is always shrewd planning.

The final step is to develop the implementation plan, including device acquisition, applications development, and user training. All of these need to be updated regarding current and future technologies, and need to be flexible in their understanding of user vernaculars (for applications development and presentation issues) and alternative device locations. A plausible hypothesis would hold that, for strategic uses within the community, optimum locations for high-end devices would not be within every home but rather in central community locations such as stores, schools, and community centers. These issues should be driven by community purposes and objectives, and not by middle-class habits.

In summary, the development of a community-based information strategy requires neither high-powered technology nor massive outside resources. On the part of the community it requires an ability to set priorities and to mobilize local resources in pursuit of those common goals. And it further requires an awareness on the part of community leadership that the world is entering into a revolution no less profound than the Industrial Revolution.

The information revolution is here. Like all revolutions its full extent will not be understood until after some of the critical forces in society have been rearranged and strategic positions established. Like all revolutions, this one is difficult to understand in the swirl of events. New forces emerge daily, and there are surprises around every corner.

Like all periods of rapid social change, this revolution represents an opportunity for disadvantaged communities. The opportunity will become a reality only if the community can master the new information environment.

APPENDIX

The South Chicago Study

In 1995 Wizdom Systems, Inc., a computer software firm located in Naperville, Illinois, conducted a 6-month study of information requirements and user interests in the inner-city neighborhood of South Chicago. This study, "Tools for Information Empowerment," was conducted under a Small Business Innovation Research contract with the National Science Foundation. The results of the study are presented in the final report (Wizdom Systems, Inc., 1995).

The South Chicago study had three objectives:

1. To evaluate the feasibility of advanced information technology for transforming a low income community.
2. To understand the barriers to this transformation.
3. To understand the local patterns of information access and usage.

The methods used by the study included open-ended interviews on information issues in the homes of low-income families, a structured survey of information budgets and usage among a sampling of the community's residents, and the conducting of classes on Internet for mothers in an elementary school.

This last activity deserves elaboration. The project team established an Internet resource using voice-grade telephone lines in the local elementary school, and a graduate student at the University of Chicago provided training on the computer and the use of the Internet. The classes were publicized within the school, and several (self-selected) mothers took the opportunity to learn about computers and the Internet.

The findings of the study can be summarized as follows:

1. Computers and computing are seen as foreign objects and activities in this low-income community.
2. Language is a nearly absolute barrier to online information resources for low-income Hispanic families.

3. Perceived lack of education, rather than actual lack of education, was an important barrier. (That is, mothers with an eighth-grade education thought that, because their education was limited, they could not use the computer.)
4. There was a strong consensus within the community regarding information needs and priorities. The top two items, reflecting interests of more than 60% of the survey respondents, were (a) information about what was happening outside the neighborhood, and (b) information on local issues including jobs, government services, retail stores, and day care.
5. There is a willingness to experiment with new information resources, once perceptual issues have been addressed.
6. Current household budgets for all media are substantial; the median household information budget was $170 per month.
7. Families are committed to their children's education, and will spend relatively large amounts of money to support children's learning. Families reported that they would be willing to spend, on average, an additional $16 per week if it would help their children's education.

The South Chicago study concluded in June 1995, with the submission of the final report to the National Science Foundation. Developments in the information industries since then—the dramatic growth in the use of the WWW and the release by Oracle Corporation of an Internet computer for less than $300—have reinforced the conclusion of the study that providing advanced information services in the inner city is a business opportunity.

REFERENCES

Mueller, Milton, & Schement, J. R. (1995) Universal Service from the bottom up: A profile of telecommunications access in Camden, New Jersey. *The Information Society, 12, 3,* 273–291.

Schement, J. R. (1995). *Tendencies and tensions of the information age: The production and distribution of information in the United States.* New Brunswick, NJ. Transaction Publishers.

Wizdom Systems, Inc. (1995) *SBIR Phase I Final Report. Tools for Information Empowerment.* Available from Wizdom Systems, Inc., 1260 Iroquois Drive, Naperville, Illinois, 60563.

7

Universal Access To Infrastructure and Information

Allen S. Hammond IV
University of Santa Clara Law School

The efforts to reshape the communications industry are in high gear. In response to the federal government and many state governments, we can expect companies to increase their earlier efforts to jockey for competitive position and new market entry by creating new applications for existing technologies, creating new alliances and services, and pressing for favorable market entry and competition policies.

The passage of national legislation to speed the building of the information superhighway is a critical point in time to assess how the legislation addresses issues of economic development, employment, educational opportunity, electronic democracy, and service delivery. For the welfare of individuals, citizens, and workers of tomorrow will be determined in significant measure by the quality of access to the means of communication and information that is afforded to most today.

According to the politicians, regulators, and captains of industry, there will be a cornucopia of new multimedia services for various businesses and residential consumers. Far more is at stake than the availability of more movies, sports, or software-defined network capabilities with amazing applications. As we reshape our communications system and the laws that govern it, we also reshape our democracy and ourselves.

There are many potential public benefits of multimedia convergence. They include increased economic competitiveness, enhanced delivery of education, the flowering of civic democracy (Hong, 1995), and enhanced medical service delivery (Ryan, 1996). For instance, studies have identified a significant economic multiplier effect that occurs as a result of a nation possessing an efficient telecommunications infrastructure (Warwick, 1993). Researchers have documented impressive learning gains achieved through the use of interactive video and computer-based instruction. In some cases, such instruction has been found to be 30% more effective than more traditional instructional formats (*Educational Technology: Hearing Before the*

Subcomm. on Labor, Health and Human Services, and Education, and Related Agencies of the Senate Committee on Appropriations, 1995). The use of electronic mail (e-mail) services, computer bulletin board systems (BBS), and computer conferencing systems as channels to make decisions and disseminate information can help grassroots political organizations, nonprofit groups, and other public interest groups to gather critical information, organize political action, sway public opinion, and guide policymaking (Snider, 1994). Radiology imaging services allow physicians at a rural hospital to quickly exchange x-rays and other medical images for consultation with radiologists at larger urban facilities.

So, as we revise our telecommunications infrastructure and access to the services and information it can provide, we are revising what it means and what it takes to be an informed and responsible citizen, a well-educated individual, and a desired, competitive worker. In short, we are revising a part of the social compact under which participation in our democracy, society, and economy will be afforded and to whom full participation will be afforded.

Although the litany of potential benefits grows ever larger and more impressive, there is another side to this picture. Too often as an afterthought, an early apology, or an alarm, some limited attention is given to potential impact of these national and increasingly global developments on inner-city urban and rural communities and the people who live there.

There is growing concern that a significant portion of America's inner-city and rural communities are in danger of becoming the domain of the "information have-nots" (Shiver, 1995). The reasons for this concern can be summarized in four words: employment, education, infrastructure, and access. First, from an employment perspective, there is an absence in inner-city communities of a well-educated, learning-oriented workforce capable of taking the high technology jobs that the National Information Infrastructure (NII)-related industries are generating. The National Information Infrastructure is composed of telephone, computer, data, television, cable television, and wireless transmission networks and the equipment attached to them. Meanwhile, the low and unskilled jobs for which many in the inner city and rural populations are suited are dwindling rapidly (Ehrenhalt, 1993). Second, attempts to address this lack of preparedness with education are undermined by high student dropout rates and poorly staffed, underequipped, underfinanced schools, as well as by the unequal spread of educational technology (*Appropriations for 1997: Hearings Before the Subcommittee on Labor, Health and Human Services, and Education, and Related Agencies of the House Committee on Appropriations*, 1996). There is also a decided absence of computer literacy in many inner-city and rural communities for a variety of reasons, such as its members' low income and lack of education.

Even where computer technology is employed, there are often disparities in the manner in which it is used. For instance, in poor inner city schools and schools with high percentages of multicultural students, students are less likely to receive assignments calling for hands-on or higher order thinking activities to help them think critically (Rockman, 1995). Instead, they are more likely to receive computer instruction for isolated skill development or remediation (drill and practice), repeating applications of basic reading and math computation skills that are then assessed via standardized tests.

The telecommunications and electronic network infrastructure serving the residents of these communities is often older and less able to provide the newer services and functionalities that middle- and upper-class neighborhoods can take for granted ("NYNEX Accepts," 1995). Moreover, the current evolution of telecommunications competition policies provide ample justification for the concern that significant portions of such communities will not be the first priority of competitive access providers, telephone, or cable companies seeking to expand and/or upgrade infrastructure to garner, protect, or increase market share. For instance, many groups hold telecommunications companies responsible for designing systems that will bypass poor, remote, or minority communities (Sheppard, 1995).

The electronic redlining, universal service, and cable must carry controversies are cases in point. Each is directly related to government efforts to formulate new competition policies that have placed pressure on the subsidies that make it possible for the majority of U.S. households to have access to a telephone (94%) and television for access to video information (98%) ("NTIA Study," 1995).

According to many, increased competition in the local telephone service area is increasing pressure to abolish, restructure, or reallocate the cost of subsidies that underwrite the provision of universal phone service to many inner-city and rural residents (Arnst & Kelley, 1995). The Communication Act of 1934, Title I, Section 1, set forth the goal of American communications policy "to make available, so far as possible, to all people of the United States a rapid, efficient, nation-wide, and world-wide wire and radio communications service with adequate facilities at reasonable charges." In telephony, this policy evolved into the requirement that monopoly telephone companies provide service to as many as possible (Weikle, 1995). The companies were allowed to subsidize the cost of serving poor, rural, or other less-profitable customers with higher margin clients such as downtown businesses (Taylor, 1995). Even with this inclusive goal of universal service in place, there are substantial gaps in service. Roughly 25% of low income families lack phone service (Ruben, 1995). Pay phones, the traditional substitute for a phone in the home, are becoming less and less accessible (Johnson & Spielman, 1994). To combat their use in facilitating drug dealing, and other illicit activities such as phone fraud and computer hacking, pay phones have been removed from many poor neighborhoods, restricted to outbound calls only, or returned to rotary dialing (Davis, 1994). Without access to a phone, poor people lose opportunities to acquire government benefits, secure employment, obtain emergency care, and have access to other essentials such as connection to the larger society. In a recently published study of obstacles that predict a lack of having a regular medical provider and delays in seeking medical care for patients at an urban public hospital, 20% of respondents cited lack of phone service (McNagny, Parker, Rask, & Williams, 1994).

Now, as competition increases in local telephone markets, substantial pressure is being generated to undo or substantially revise the universal service subsidy structure (*National Communications Infrastructure (Part 3): Hearings on H.R. 3636 Before the Subcommittee on Telecommunications and Finance of the House Committee on Energy and Commerce*, 1994). As mentioned before, a good portion of the underwriting of basic service takes the form of subsidies (*Competition at the Local Loop*, 1993). There are direct subsidies that help telephone companies with higher than average cost areas ("New Study," 1995). These companies operate in rural areas where telephone lines must be strung for many miles to reach few subscribers. There are also cross subsidies in which one group of users pays higher prices to underwrite lower prices for other groups. These include: business-to-residential customers; urban-to-rural subscribers; and long-distance-to-local callers.

Competition and changing regulatory attitudes are challenging the subsidy structure. For instance, long-distance telephone companies seek a reduction in access charges that comprise approximately 40% to 45% of the cost of long-distance charges (Bray, 1995). Efforts to reduce costs include bypass arrangements and efforts to enter the local market as competitors. Meanwhile, long distance companies are joined by competitive access providers (CAPs) who compete for the high end users that comprise a small number of actual customers but generate the vast majority of revenues (Arellano, 1995). Finally, consumers want the benefit of any reductions in cost of service due to increased efficiencies in the network but do not want to subsidize local telephone company entry into competitive cable and information services from which the telephone companies' competitors are emerging (*Competition at the Local Loop*, 1993). Thus revenues for competitive entry or introduction of competitive services must come from local exchange carriers' (LEC) profits garnered via the provision of competitive services and from shareholders. The net result is pressure on whatever subsidies exist to fund universal service.

Competition and innovation in the broadcast and cable markets is creating a shift from advertiser-supported to subscription-supported delivery of video information (Artzt, 1994). This shift has prompted congressional action to protect the information access of approximately 40% of the U.S. public by enacting the cable must carry laws, which require cable operators to carry a number of commercial and public broadcast stations operating in the operator's franchise area. The Supreme Court held recently that the government had a compelling interest in assuring that 40% of the country retained access to "free" broadcast information (*Turner Broadcasting Systems v. FCC*, 1994).

In addition, those who rely on subsidized access to information provided by public libraries may soon find their access diminished as well as telecommunications firms and online providers supply subscription-based alternatives to ofttimes limited library holdings (Young, 1994).

Recently a coalition of consumer and civil rights groups alleged that several regional telephone companies were bypassing low-income and minority communities in the early stages of building new video dial tone networks ("Groups Petition," 1994). These networks would allow their traditional phone customers to order movies, TV shows, sportscasts, and other programming from computer databases that would act like giant electronic video stores. Because these video dial tone networks could become the primary communications system for millions of Americans, it is argued they must be made available in an equitable and nondiscriminatory manner. Many believe that without safeguards, state procompetition policies will actually encourage redlining (Moore & Polanco, 1995).

Under the recently enacted legislation, video dial tone networks are called open video systems and the telephone companies expected to provide them are given greater latitude in entering the video distribution market ("Open Video," 1996). Regardless of their name, the short-term economics are unlikely to change—systems will be built first in communities that possess the requisite dollars and demand (Carlini, 1995). Many inner-city and rural communities as well as near suburbs are not viewed as desirable markets regardless of their actual consumption of telecom and video services. When the marketplace does not work to distribute goods and services equitably, we call it a *market failure*.

How does the legislation address the service and information access questions issues raised? As the federal government unleashes competition, it has made provision for the protection of universal access (Telecommunications Act of 1996, § 254). For this, the Congress and the President, as well as all those people of good will who fought for such provisions, are to be commended.

Pursuant to law, the development and implementation of universal service is to be governed by a set of guiding principles that articulate a general policy of equitable distribution of reasonably priced, affordable, quality services. The Federal Communications Commission (FCC) will assure that the rates charged consumers are just, reasonable, and affordable. Those Americans who are deemed poor by established criteria are protected from a loss of service as well. The FCC must promulgate and implement regulatory policies that assure communities, schools, libraries, and health care facilities access to telecommunications services and advanced telecommunications services. These services are to be provided at a discount that has not yet been made available.

Section 254 does not identify the service which must be universally available. Instead, it requires the FCC in conjunction with representatives of the state regulatory bodies to create a joint board charged with making recommendations about which services to include. The FCC recently adopted the Joint Board's proposed definition. Currently "basic" telephone services include touch tone, single party, operator assistance, directory assistance, and emergency 911 services. Access to e-mail, the Internet, and other multi-media services is not required. The FCC must respond to subsequent Joint Board recommendations within 1 year after receipt and must review the progress of the implementation of universal service within 5 years of the policy's implementation.

Other critical requirements implementing the national policy of equitable access are that telecommunications service providers are required to make an equitable and nondiscriminatory contribution to the preservation and advancement of universal service. All federal and state mechanisms for funding and implementing universal service must be specific, predictable, and sufficient to preserve and advance universal service. And, elementary and secondary schools and classrooms, health care providers, and libraries should have access to advanced telecommunications services.

The FCC and the Joint Board are also given the flexibility to establish additional principles as the commission and the joint board determine are necessary and appropriate for the protection of the public interest, convenience, and necessity consistent with the Telecommunications Act. Changes to the definition of universal service may be made based on a recognition of the evolving nature of telecommunications services consistent with: advances in telecommunications and information technologies and services; a recognition that telecommunications and information technologies and services are essential to education, public health, or public safety; the operation of market choices by customers such that the additional services have been subscribed to by a substantial majority of residential customers; additional services are being deployed in public telecommunications networks by telecommunications carriers; and the addition of services are consistent with the public interest, convenience, and necessity.

Should the service roll-out be deemed too slow or inequitable, the FCC may stimulate faster deployment of services, provided the FCC does so by facilitating more competition. Of course, the FCC is faced with a market failure problem. Due to misperceptions or accurate perceptions of a market segment's desirability, more competition is not likely to solve it.

In addition to wiring in the schools as special recipients of universal access, the law establishes the National Education Technology Funding Corporation to provide loans and grants to educational institutions and libraries to fund access to telecommunications networks and to technology-based educational tools and innovations (Telecommunications Act of 1996, § 708). However, anyone watching the budgetary wars in Congress and the state legislatures knows that the pool of federal and state education dollars is likely to shrink rather than grow (Campbell & Fischel, 1996). Under such circumstances, the law's effectiveness will be qualified by future allocation decisions, as well as agency budgetary considerations and the priorities and efficiency with which state bureaucracies distribute the federal largesse.

While the law directs the Joint Board and the FCC to address the question of universal access, it does not begin to confront the question of how to ensure continued access to information for all Americans in an era of increasing cost.

The cost of access to information is increasing as the equipment necessary for reception shifts from a television set to a television set, a set top converter (cable) and a computer, even as access to the information requires a monthly fee and possibly a per-minute charge (Shapiro, 1995). As subsidy mechanisms such as advertiser-supported broadcast programming are replaced by subscription-based cable and computer access to programming, access to technology will not ensure that there is valuable programming information available at reasonable rates as well.

Despite early efforts to include specific language prohibiting the electronic redlining of poor and minority communities in the earlier drafts of legislation, the Act does not address electronic redlining directly. Instead, the Act includes an amendment to Section 151, which prohibits discrimination in the provision of service on the basis of race, color, religion, national origin, or sex. Although clearly an important amendment, it fails to address the minority and poor communities' concern that they are being less economically desirable, not necessarily because they are composed of religious, ethnic or racial Americans, or women. Given the current split in eligibility based on whether the services at issue are basic or advanced, many Americans are not likely to receive access to all services immediately absent the ability to pay or to be seen as an economically desirable market. As a result, even with nondiscrimination principle in Section 151, and, even if the access to "basic" services is actually achieved via Section 254, the Telecommunications Act does not prohibit or discourage business decisions to delay or decline to provide new services in many communities.[1] So long as the competitive business decisions are based on perceptions that communi-

---

[1]Ted Hearn, Ameritech: We Don't Redline, *Multichannel News*, July 29, 1996, p. 3; John J. Oslund, U.S. West's Long March, *Star Tribune*, July 17, 1995, p. 1D; Ted Hearn, Cable Accuses Bells of Targeting Only Wealthy, *Multichannel News*, April 17, 1995, p 18; Ed Rose, The haves and the have-nots; access to information Technologies, *Communication World*, November, 1994, p. 22.

ties possess limited wealth, or constitute less desirable markets, those communities will continue to be left out for the near term when it comes to the provision of all services not deemed "basic."

Being a baby boomer, my life spans the beginning of the era in which electronic media overtook the printed word and the era in which the Internet and open networks may overtake the television, telephone, and cable. During this time, there have been a plethora of enduring images. At least three seem appropriate to mention today.

Two are images from my childhood education and come from books. The third is uniquely electronic. The first is a picture of Thomas Jefferson writing the Declaration of Independence, giving voice to the U.S. version of the Enlightenment ideal "that all men are created equal." The second is a picture of a young Abraham Lincoln reading a book by candlelight, with its strong message about the transformative value of the self-motivated pursuit of knowledge. From my adulthood comes the third image of a recent president, Ronald Reagan, whose mantra of marketplace economics and self-interested consumerism has become the balancing cornerstone to the "New Deal" of FDR and the "Great Society" of JFK and LBJ.

For me, these images are metaphors for the three roles that define who we are as Americans. From the democratic vision of Jefferson, our enduring sense of equality and responsibility as citizens in a democracy; from Lincoln, our evolving, self-empowering role as lifelong learners; and from Reagan, our concurrent role as consumer drivers of the economic engines of our economy.

They are also the three roles that are potentially at war when we consider this new law. Our goal is to facilitate learning and electronic democracy as well as economic competitiveness. But, we are told that it is as consumers that we usher in this new era; that as consumers we may exercise our economic franchise and determine what equipment, technologies, and services will be offered, at what prices, quantities, and to whom. There is a major flaw in this prescription. Our nation has the greatest discrepancy between the economic classes of all major industrialized nations (Bradsher, 1995). To the extent that dollars of disposable income drive deployment, we have written a prescription for market-driven democracy. I submit that such a prescription is fraught with potential for disaster for those already on the verge of exclusion because of their economic polity. Unlike economic goods, however, democracy, citizenship and individual freedom should not be distributed on a free market, pay-as-you-go basis.

Without a continuous quest for and acquisition of knowledge, we cannot be well-informed citizens or consumers. Without the exercise of informed citizenship, we risk dimming the light of knowledge, muzzling freedom of speech, and undermining the strength of our economic polity. Without informed consumerism, we run the risk of exercising our economic franchise in ways that diminish the value of citizenship and consign many to ignorance.

The universal service provisions of the bill are an admirable first step in addressing these risks. However, the work in assuring universal access to self-empowering telecommunications technologies and to information they can provide has just begun. Proceedings will eventually begin in every state as well as with the FCC to define what constitutes universal service and how to pay for it. As these proceeding occur, we must assure that all Americans are represented and that the guiding principles of the congress indeed provide the floor past which no American can fall.

ACKNOWLEDGMENT

Special thanks are due to my research assistant Nancy Bloom who assisted in the research and subsequent editing of this manuscript.

REFERENCES

Appropriations for 1997: Hearings Before the Subcommittee on Labor, Health and Human Services, and Education, and Related Agencies of the House Committee on Appropriations, 104th Cong., 2d Sess. 309, 309 (1996) (statement by Sharon P. Robinson, Assistant Secretary for Educational Research and Improvement, Department of Education).

Arellano, M. (1995, July). Exploiting the LECs' achilles' heel. *Telecom Strategy Letter*, 81.

Arnst, C., & Kelley, K. (1995, February 20). Phone frenzy: Is there anyone who doesn't want to be a telecomm player? *Business Week*, 92.

Artzt, E. L. (1994, May 23). P&G's artzt: TV advertising in danger; remedy is to embrace technology and return to program ownership. *Advertising Age*, 24.

Bradsher, I. (1995, August 20). Ideas & trends: More on the wealth of nations. *The New York Times*, § 4, p. 6.

Bray, H. (1995, October 16). Bill limits or protects, depending on biz size. *Crain's Detroit Business*, p. 32.

Campbell, C. D., & Fischel, W. A. (1996, March). Preferences for school finance systems: Voters versus judges. *National Tax Journal*, *49*, 1.

Carlini, J. (1995, June 19). Universal service-or regulated "telewelfare"? *Network World*, 51.

Communication Act of 1934, 47 U.S.C. § 151.

Competition at the local loop: Policies and implications. Forum report of the 7th annual Aspen conference on Telecom Policy, Aspen, Colorado. (1993, March 15). *Business Wire*.

Davis, R. (1994, July 13). New stab at lifeline of dealers; pay phones face 2nd Daley attack. *Chicago Tribune*, p. 1.

Educational Technology: Hearing Before the Subcomm. on Labor, Health and Human Services, and Education, and Related Agencies of the Senate Committee on Appropriations, 104 Cong., 1st Sess. 5,6 (1995) (statement of Madeline M. Kunin, Deputy Secretary, Department of Education).

Ehrenhalt, S. (1993, February). Economic and demographic change: The case of New York City. *Monthly Labor Review*, *116*, 40.

Groups petition FCC for prohibition of "electronic redlining" by RHCs. (1994, June 1). *Advanced Intelligent Network News*, *4*.

Hong, P. (1995, February 26). Losing the cyberspace race: On the fast-paced high-technology track, inner city schools and libraries are taking small steps to catch up. *Los Angeles Times*, p. 12.

Johnson, M. A., & Spielman, F. (1994, July 13). Daley: Ban pay phones from private property. *Chicago Sun-Times*, p. 4.

McNagny, S. E., Parker, R. M., Rask, K. J., & Williams, M. V. (1994, June 22). Obstacles predicting lack of a regular provider and delays in seeking care for patients at an urban public hospital. *The Journal of the America Medical Association, 271,* 1931.

Moore, G., & Polanco, R. (1995, July 3). Avoiding a new era of redlining. *The San Francisco Chronicle,* p. A17.

National Communications Infrastructure (Part 3): Hearings on H.R. 3636 Before the Subcommittee on Telecommunications and Finance of the House Committee on Energy and Commerce, 103d Cong., 2d Sess. 232–68 (1994) (testimony of Ivan J. Seidenberg, Vice-Chairman of NYNEX Corp.).

New study says target, don't expand, universal service subsidies. (1995, January 26). *State Telephone Regulation Report, 13.*

NTIA study sees info age "have-nots" as poor, young, uneducated. (1995, August 21). *State Telephone Regulation Report, 4.*

NYNEX accepts plan to freeze phone rates. (1995, July 4). *The Buffalo News,* p. 7B.

Open video rules in watched closely telcos hail OVS as successor to burdensome video dial tone. (1996, April 15). *Interactive Video News, 4.*

Prepared statement of Kevin O'Brien vice president and general manager KTVU-TV, channel 2 San Francisco, CA and chairman of the board the association of local television stations before the house commerce committee. (1996, March 21). *Federal News Service.*

Rockman, S. (1995, June). In school or out: Technology, equity and the future of our kids. *Communications of the ACM, 38,* 25.

Ruben, B. (1995, September 22). Access denied: Inequality in access to information networks. *Environmental Action Magazine, 27,* 18.

Ryan, M. (1996, April 29). "Distance health care" is latest medicine. *Electronic Engineering Times, 899,* 55.

Sheppard, N. (1995, July 4). Internet in the classroom: High-tech push fights electronic redlining. *Chicago Tribune,* p. 1.

Shiver, J. (1995, March 29). Busting barriers to cyberspace. *Los Angeles Times,* p. A1.

Snider, J. H. (1994, September). Democracy on-line: Tomorrow's electronic electorate. *The Futurist, 28,* 15.

Taylor, R. (1995, February 21). Leveling the field in telecommunications. *St. Louis Post-Dispatch,* p. 11B.

Telecommunications Act of 1996, Pub. L. No. 104-104, 110 Stat. 56, § 254.

Telecommunications Act of 1996, Pub. L. No. 104-104, 110 Stat. 56, § 708.

Turner Broadcasting Sys. v. FCC, 114 S. Ct. 2445, 2453–54 (1994) (citing 47 U.S.C. §§ 531, 534(b)(1)(B), requiring cable systems with more than 12 active channels to set aside one third of their channel capacities for commercial broadcasters requesting carriage; 534(b)(1)(A), requiring cable systems with more than 300 subscribers but 12 or fewer active channels to carry signals of three commercial broadcasters; and 535(a), requiring carriage of public broadcast stations).

Warwick, M. (1993, July). Rural comms: Wishful thinking versus reality; rural communications. *Communications International, 20,* 44.

Weikle, J. (1995, March/April). Ready for prime time—universal service meets universal competition. *Rural Telecommunications, 14,* 50–53.

Young, P. R. (1994, June). Changing information access economics: New roles for libraries and librarians. *Information Technology and Libraries, 13,* 103.

IV

PAYING FOR UNIVERSAL SERVICE

8

Overview of Universal Service

Carol Weinhaus
University of Florida

Robert K. Lock
Illinois Commerce Commission

Harry Albright
Ameritech

Mark Jamison
Sprint

Fred Hedemark
AT&T

Dan Harris
Bell Atlantic

Sandra Makeeff
Iowa State Utilities Board

Currently there are debates over the provision of telecommunications services in the United States. The debates involve a number of issues; however, one of the most important is the debate over universal service. Although technically undefined, universal service is generally interpreted to mean the widespread availability of telecommunications service to all Americans at affordable rates. The debates center around a number of questions:

What should the definition of universal service be?

Are subsidies needed to keep universal service "affordable?"

What is the appropriate level of subsidy?

Who should pay for the subsidies?

How do you insure that "basic local service" remains affordable in an environment of increasing competition?

The objective of this chapter is to provide an overview of the work that the Telecommunications Industries Analysis Project (TIAP) has done on the subject of universal service between 1991 and 1995. This chapter uses key figures and text from TIAP research in a summary format to explain the issues, quantify the current subsidies, and lay out various policy options.

The rest of this chapter covers the following items:

Current Subsidies: Gives estimates for the amount of various current subsidies and explains why current subsidies are needed. Also shows how some customers subsidize other customers.

A Need for New Subsidy Structures: Explains how the current subsidy structure is incompatible with the increasingly competitive communications marketplace. Also shows the potential impact of eliminating one form of subsidy, rate averaging, on rural customers.

A New Basis for Subsidies: Shows how existing subsidy structures can be changed to work in both competitive and noncompetitive markets. Also explains the need for transitions from the current system to a new system and explains how that can be done.

Implications of Changing What Universal Service Means: Gives the impacts of changing the definition of universal service to include advanced telecommunications services and of mandating service before it becomes widely available.

CURRENT SUBSIDIES

The concept of universal service has long been a part of the telecommunications industry in the United States. However, over time its meaning has changed. Originally, the concept of universal service meant a widespread, interconnected telephone network wherein every user of the network could connect with every other user (Mueller, 1993). Theodore Vail, President of AT&T, first coined the term "universal service" in 1907. The 1910 *Annual Report* of AT&T contains the following statement: "[The Bell System] believes that the telephone system should be universal, interdependent and intercommunicating, affording opportunity for any subscriber to any exchange to communicate with any other subscriber of any other exchange within the limits of speaking distance" (see Mueller, 1993). Although this definition may seem identical to today's definition, it did not embody the system of subsidies designed to assure affordable service to everyone. The idea was simply that the network became more valuable as a whole as more people were added to it.[1]

Recently, the concept of universal service has evolved to mean the provision of service to everyone at affordable rates. With that redefinition has come a complex system of subsidies wherein some companies and customers pay a portion of the costs of providing service to other customers.

[1]For a more complete discussion of the development of universal service, see Muller (1993).

Within the traditional telephone industry, these subsidies are inextricably intertwined with one another, and with the cost accounting structure regulating the entire industry. These subsidies (both explicit and implicit) are generally designed to achieve the public policy goals of assisting low-income households and keeping basic local service rates affordable.

Figure 8.1 lists many of the subsidies embedded in the current system. Although this system worked fairly well in a monopoly environment, growing competition is rendering this system obsolete.

What Are the Amounts of the Subsidies?

There are numerous studies that attempt to quantify the various subsidy amounts. Figure 8.2, shows the questions asked by each study and the amount of the subsidy estimated. The amounts from each study are different because each study asks different questions. There is a temptation to add or subtract these different study results; however, without detailed knowledge of each study, the resulting calculations are meaningless.

Why Are the Subsidy Amounts Different?

In addition to answering different questions, each study uses different approaches, definitions, data sets, and subsidy flows. For example, Fig. 8.3, shows that at any particular time, subsidies flow along three directions: between services, between geographic areas, and between customers. In addition, there are various views as to what constitutes incremental cost. For example, the MCI/Hatfield and the USTA/Monson-Rohlfs studies each use different definitions.

Why Are Subsidies Needed?

Figure 8.4 illustrates one reason why subsidies are needed. In this case the subsidy is used to help cover the cost of the local loop (see Fig. 8.1). Figure 8.4 shows that regardless of the method for defining loop costs (embedded, embedded without overheads, proxy, or future costs), in some cases the price for local service does not cover its cost. Furthermore because the prices have to cover more than just the cost of the loop, this figure understates the issue. For example, the costs for switching and operator services are not included in this figure. The difference between costs and rates may be much greater due to customer location and the density of customers in an area.

The residential and business rates indicate nationwide averages.[2] Because the business rates in most cases are sufficient to cover the average embedded cost of the loop, most of the local subsidies are for residential customers.

---

[2]In the traditional telephone industry, business rates are generally higher than residential rates.

Rate and Cost Averaging to Achieve Public Policy Goals:
Provide "reasonable" rates on a nondiscriminatory basis.
Allocation of historic costs to determine costs for pricing.
Differences between business and residential rates.
Use of fully distributed cost methodology to allocate common overheads.
Pricing averaged across broad geographic areas to promote universal service and infrastructure development.
Etc.

Financial Assistance to Ensure Universal Service:
Targeted high-cost and low-density areas:
- High Cost Fund (HCF).
  - High Cost Loop Fund, formerly called Universal Service Fund (USF).
  - Long-term support (LTS).
  - Small telephone company local switch support, or Local Switching Support.
- REA loans.

Low income households:
Low Income Fund, consists of Lifeline/Link-up programs.
Schools, libraries, and rural health care providers.
Offshore areas:
- Assistance to Alaska, Guam, Hawaii, Micronesia, Puerto Rico, and the Virgin Islands, for interconnection to traditional industry network in the contiguous 48 states.

Obligation to Serve:
Carrier of Last Resort.
- High-Cost and Low-Cost Locations.
- Facilities ready to serve a customer (large or small) whenever they want.
- Interconnection (Mobile Carriers, Competitive Access Carriers, Providers, etc.).

Intrastate Services Paying 75% of Shared Local Facility Costs:
Local Exchange Rates.
Interexchange Usage Charges.
Other.

Special Needs Assistance for Equivalent Access to Telecommunications Network:
Telecommunications services for hearing-impaired and speech-impaired individuals.

Oversight of Jurisdictional Shifts:
Participation through Federal-State Joint Board.
Maintain "reasonable" basic local service rates.

Depreciation Policies:
Multiple Mechanisms and Authorities.

FIG. 8.1. What are the subsidies? Current public policy goals: some support mechanisms and forms of averaging associated with the regulated telecommunications industry. From Weinhaus et al., 1992, *Who pays whom: Cash flow for some support mechanisms and potential modeling of alternative telecommunications policies*, p. 4. Copyright 1998 by Carol Weinhaus and the Telecommunications Industries Analysis Project Work Group. Boston, MA.

How Have Interstate Long-Distance Customers Subsidized Local Customers?

Figure 8.5 shows the 1992 cash flow for payments made by long-distance customers to support local customers, including interstate subsidies. The shaded box indicates the amount of long-distance company revenues that are paid to local telephone companies for interconnection and for support. Approximately 45% of the long-distance companies' costs are payments to

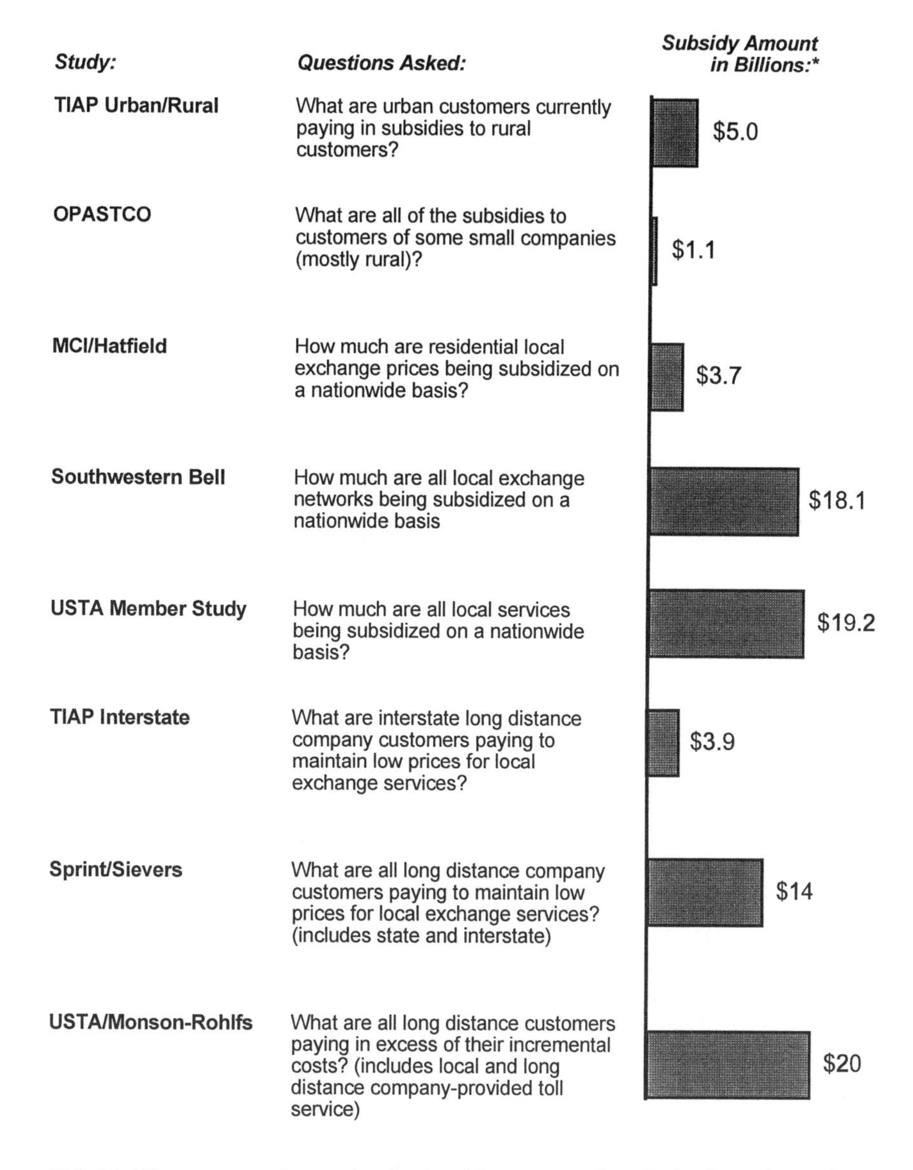

FIG. 8.2. What question does each subsidy address? Examples of subsidy studies and their results. From Weinhaus, Pitts, Jamison et al., 1994, *Apples and oranges: Differences between various subsidy studies*, p. 2. Copyright 1994 by Carol Weinhaus and the Telecommunications Industries Analysis Project. Reprinted with permission.

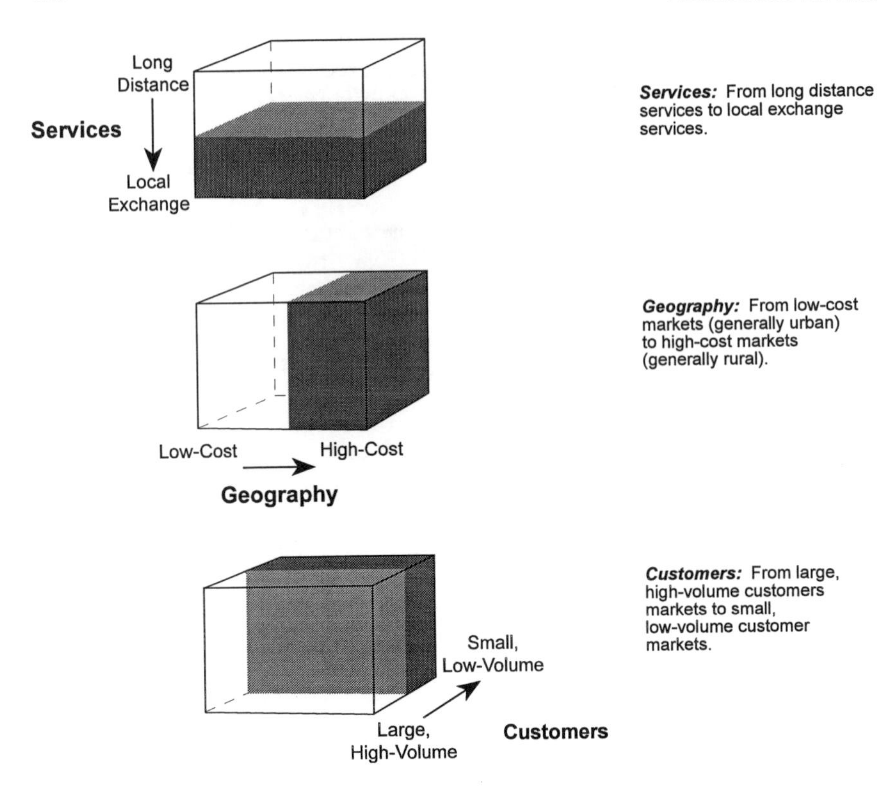

FIG. 8.3. Why are the subsidy amounts different? Subsidy flows by three market segments. From Weinhaus, Pitts, Jamison et al., 1994, *Apples and oranges: Differences between various subsidy studies*, p. 2. Copyright 1994 by Carol Weinhaus and the Telecommunications Industries Analysis Project. Reprinted with permission.

local exchange companies. These support payments reflect public policies that include assistance to high-cost areas, low-income households, and to keep basic local service prices low.

This is a traditional view of the communications markets. This view assumes that traditional local and long-distance companies are the only players. Limiting the discussion of subsidies to these two types of companies reflects the old boundaries that defined the traditional telephone industry. Interstate and other subsidy mechanisms created a payment system to promote things people wanted.

The current subsidy mechanisms are inconsistent with the new competitive environment. Customers and companies who pay subsidies look for ways to avoid them. Existing subsidy payments benefit some companies more than others. The companies who benefit the most, therefore, have a financial stake in preserving the current system.

A NEED FOR NEW SUBSIDY STRUCTURES

Why Is a Change in the Subsidy Structure Needed?

Figure 8.6 shows that although existing government policies still treat various communications industries along traditional industry boundaries,

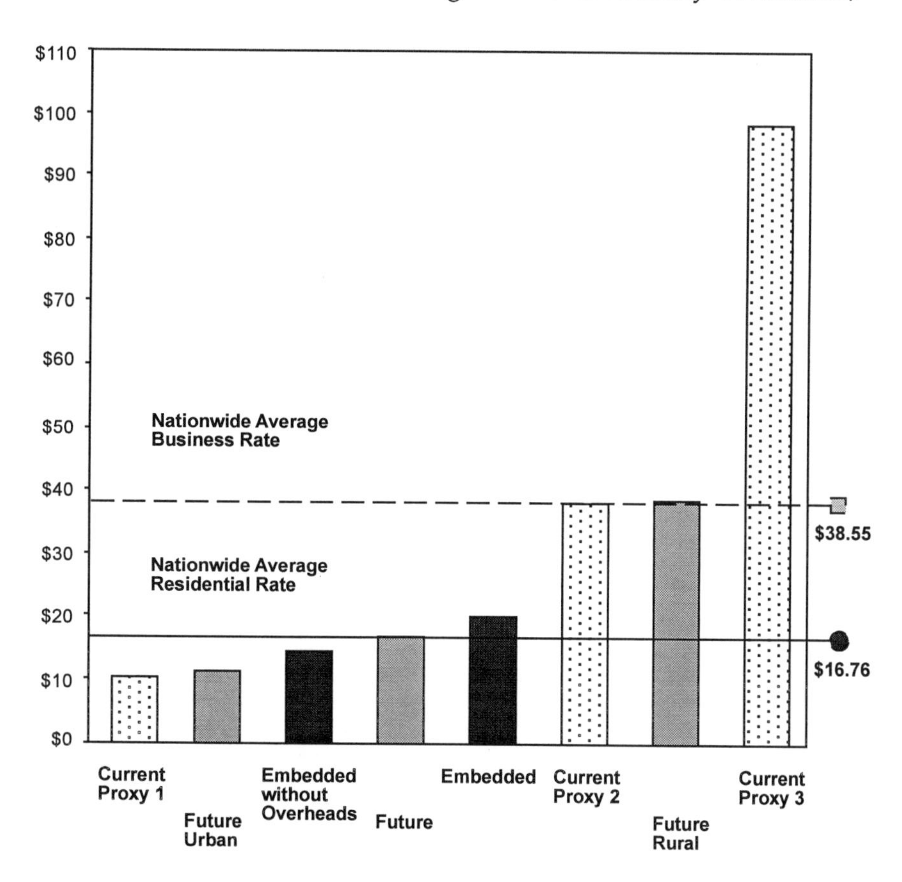

FIG. 8.4. Why are some subsidies needed? Comparison of 1993 national average residential and business rates with loop costs. Embedded costs are from the *FCC Monitoring Report*, Table 3.3; current proxy costs were developed by applying long-run different population density and distance from the telephone company's central office; future costs (future urban, future rural, future nationwide) are from the TIAP *New Technology Deployment Model* adjusted to allow comparisons with other methods for determining loop costs. For more details, see Weinhaus et al. (1995). *Loop Dreams: The Price of Connection for Local Service Competition*. Copyright 1995 by Carol Weinhaus and the Telecommunications Industries Analysis Project. Reprinted with permission.

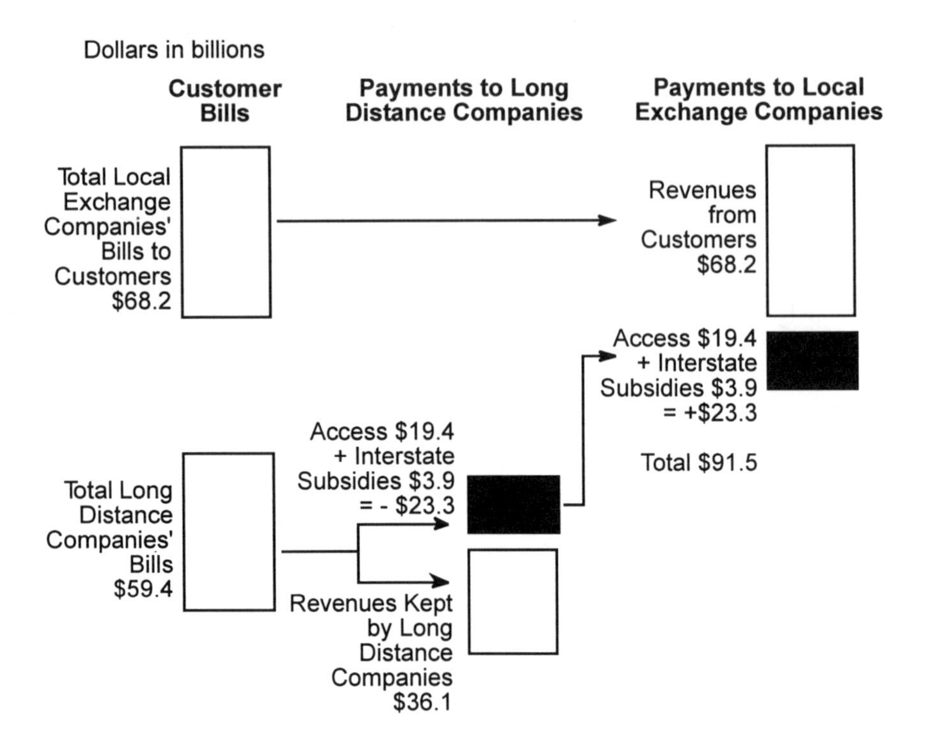

FIG. 8.5. What is the amount of subsidy flow between traditional interstate long distance and local markets? 1992 cash flow for some support mechanisms: explicit subsidies. Subsidies shown are explicit interstate subsidies only. Access is state and interstate. Adapted from models used in Weinhaus et al. (1992). Data from Federal Communications Commission (1991/1992) and United States Telephone Association (1992). Copyright 1995 by Carol Weinhaus and the Telecommunications Industries Analysis Project. Reprinted with permission.

new technologies and competition are making these boundaries obsolete. Different public policies were developed for each traditional communications industry publishing, telecommunications, broadcast/cable TV, and computer. These industries are no longer separate.

Traditional communications companies developed within the industry boundaries determined by technology and, in some instances, reinforced by public policy. Today, companies are crossing traditional industry boundaries to form new hybrids. These companies use technologies that allow products and services to become increasingly interchangeable.

Public policy, however, continues to treat each industry's products and services differently, depending on the traditional regulation of the industry. In some instances, public policies create artificial barriers between industries.

In the past, public policies for telecommunications created subsidy flows between various markets and used nationwide averaging as a mechanism to support these flows. The introduction of new technologies and competition has eroded the foundations for these subsidies.

What Is the Impact of Deaveraging on Rural Customers?

Traditional telephone practices keep prices low for rural customers because urban customers pay for a portion of the cost of rural telephone service. This is called *price averaging*—prices are averaged across areas with higher costs and areas with lower costs. Averaging was possible because there was no competition.

Competition is driving deaveraging of urban and rural prices. Competitors tend to pick low-cost areas and high-volume customers because they can offer service to these areas and customers at a lower price. These tend to be urban areas or the center of town in a rural area. Customers in these areas are beginning to have competitive alternatives that allow them to avoid the higher prices caused by the averaging process. The remaining customers with higher costs pay higher prices.

Figure 8.7 shows a potential impact of deaveraging nationwide urban and rural rates. In this example, prices paid by rural customers increase to cover the costs of rural telephone service. If this occurs, and if costs are unaf-

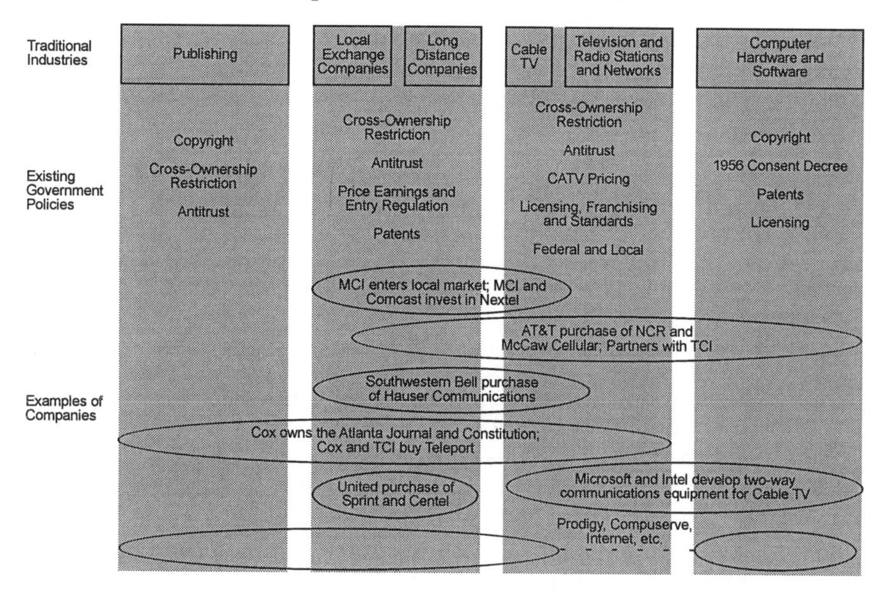

FIG. 8.6. Why is a change in the subsidy structure needed? Existing government policies, traditional industries, new alliances, 1993. From Weinhaus, Pitts, McMillin et al., 1994, *Abort, retry, fail? The need for new communications policies*, p. 6. Copyright 1994 by Carol Weinhaus and the Telecommunications Industries Analysis Project. Reprinted with permission.

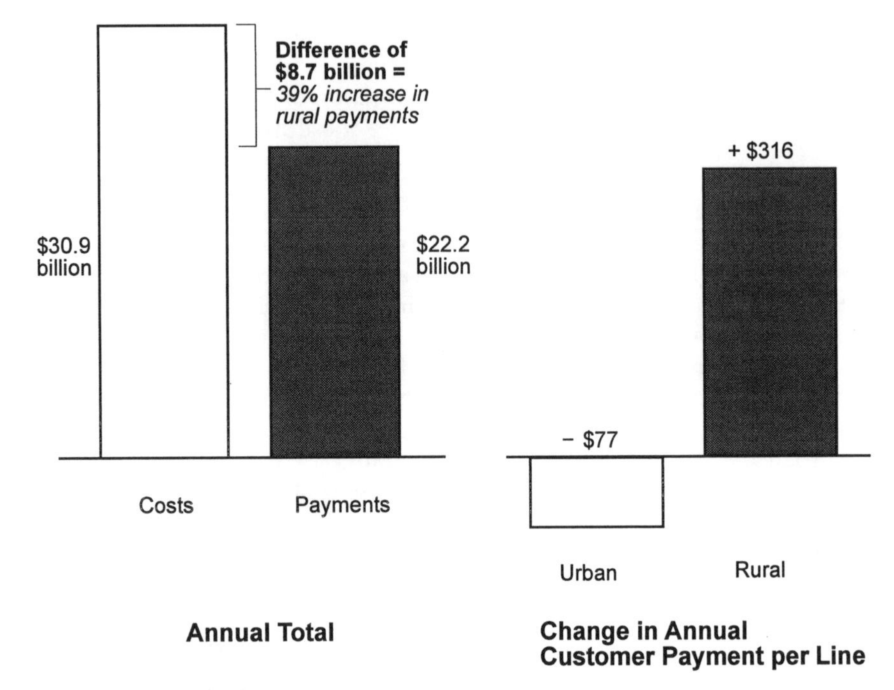

FIG. 8.7. What is the impact of deaveraging nationwide urban/rural rates? 1991 potential impact of deaveraging urban and rural rates. From Weinhaus et al., 1993, *What is the price of universal service? Impact of deaveraging nationwide urban/rural rates*. Copyright 1993 by Carol Weinhaus and the Telecommunications Industries Analysis Project. Adapted with permission.

fected by competition, then prices in rural areas would increase $8.7 billion, or 39% (1991 dollars). This translates into a decrease in urban customers' annual payments of $77 and an increase in rural customers' annual payments of $316.

What Is the Impact of Rate Deaveraging on Residential Customer Budgets?

The pie charts in Fig. 8.8 indicate nationwide average budgets for urban and rural households. The 1991 urban average budget was $31,051 and the rural average budget was $25,042. In 1991, rural customers paid approximately 2.4% of their household budgets on telephone service; urban customers paid approximately 2.0%.

The slice from each pie chart indicates the impact of deaveraging telephone rates: urban rates decrease to 1.8% of the budget and rural rates increase to 3.7% of the budget. This is less than what households spend on dining out.[3]

[3]Households spend an average of $461/year on eating out for lunch. Source: *How We Spend Our Money*, Consumer Research Center, The Conference Board. (Based on a survey by the U.S. Department of Labor, Bureau of Labor Statistics, 1995.)

What Is the Impact of Rate Deaveraging on Rural Telephone Subscribership?

Although Fig. 8.9 indicates that on average the changes are not major when taken on an individual basis, there are some customers who will be adversely affected by rate deaveraging. (See Appendix A for background information and assumptions used for Fig. 8.9)

Currently, some customers are able to purchase telephone service only because their rates are supported through averaging and other support mechanisms. Other customers might be able to pay the increased price but may choose not to because, in their judgment, the service is not worth the increased cost. Figure 8.9 indicates that current customer bills ($22.2 billion) for rural areas are supplemented with additional revenues ($8.7 billion). Therefore, the total cost of providing nationwide rural telephone service is $30.9 billion.

If the extreme case for urban and rural deaveraging were to occur and if support mechanisms were to disappear, prices for rural services would increase. Given these two assumptions, approximately 7.3% of the rural households could either no longer afford residential service or choose not to pay the higher price although they could afford it.

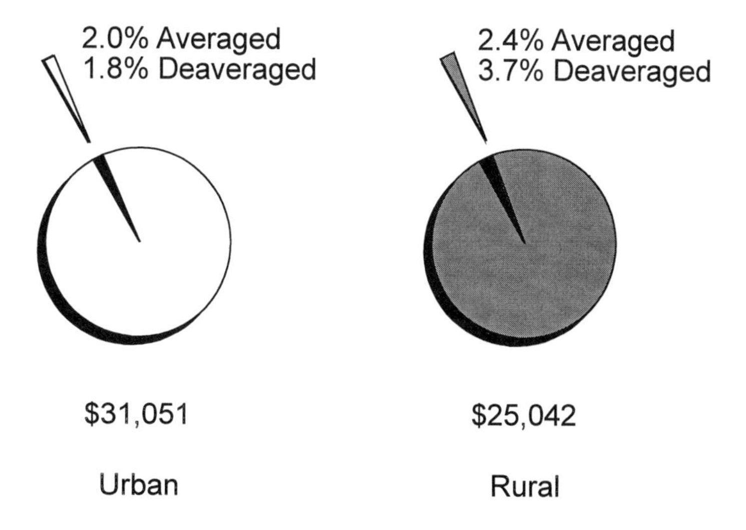

FIG. 8.8. What is the impact of urban/rural telephone rate deaveraging on residential customer budgets? From Weinhaus et al., 1993, *What is the price of universal service? Impact of deaveraging nationwide urban/rural rates.* Copyright 1993 by Carol Weinhaus and the Telecommunications Industries Analysis Project. Reprinted with permission.

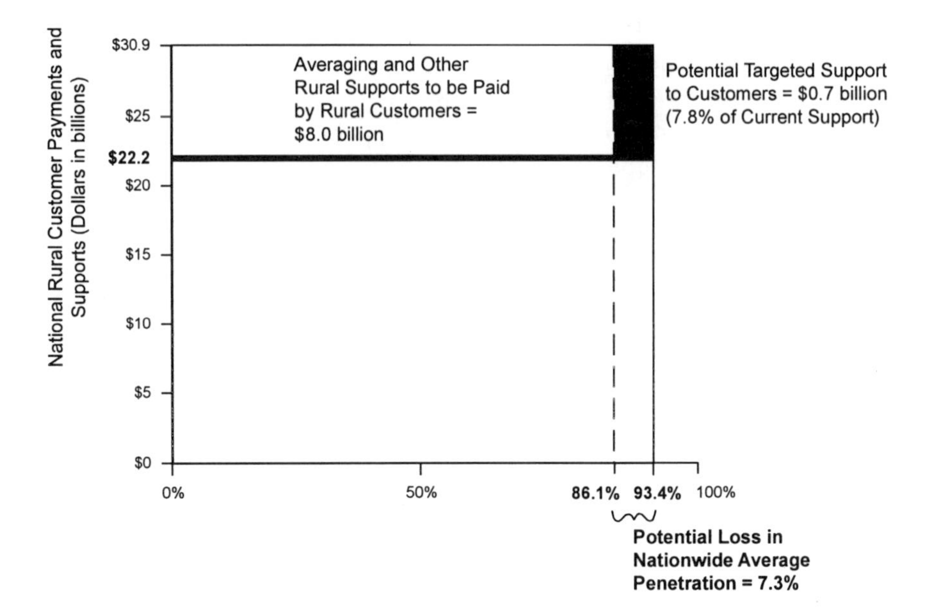

FIG. 8.9. What is the impact of urban/rural rate deaveraging on rural telephone service penetration? From Weinhaus et al., 1993, *What is the price of universal service? Impact of deaveraging nationwide urban/rural rates*. Copyright 1993 by Carol Weinhaus and the Telecommunications Industries Analysis Project. Reprinted with permission.

The vertical dashed line in Fig. 8.9 indicates the lowered penetration level. Those customers to the left of this dashed line would still buy telephone service. The prices these customers pay increase to cover the previous support ($8.0 billion). However, those households to the right of the dashed line may need additional support from external sources ($0.7 billion) if these households are to keep residential service.

This alternative answers two questions: the level of rural support and who receives it. This example assumes that 92.7% of the 1991 rural customers would pay an additional $8.0 billion in increased rates for all their telephone services. In this alternative, support of $0.7 billion is paid to those rural customers who may no longer be able to afford service or who may choose not to buy it. This is in contrast to current policies that direct most support payments to the high-cost companies and their customers, regardless of the customers' ability to pay for the service.

How Can Subsidies Accommodate Both Competitive and Noncompetitive Markets?

Figure 8.10 illustrates one approach to determining the amount of subsidy needed. This approach is called the benchmark subsidy method. In this approach, a single mechanism applies for subsidies regardless of whether a market is competitive. The benchmark subsidy method provides incentives to service providers to be efficient and allows competitive markets to operate efficiently.

Under the Benchmark Subsidy Method, a benchmark price would be determined for telephone service in an area. The regulatory process would then determine the new subsidy amount for an area by subtracting the benchmark prices from the costs. Subsidies would not be necessary when benchmark prices were equal to or above costs. Costs may be determined by any of a number of methods.

Subsidies could be general, specific, or both. An example of a general subsidy is one applied to all customers in an area. An example of a specific subsidy is one applied to low-income customers. The chart illustrates both of these examples. In the same area, it is possible to have both of these subsidies. The dashed horizontal lines indicate two different benchmark prices.

How Do You Transition From the Current Environment to the Future?

Although it is necessary to look at long-term revisions to the current structure, a transition process is necessary to get us from here to there. Transitions

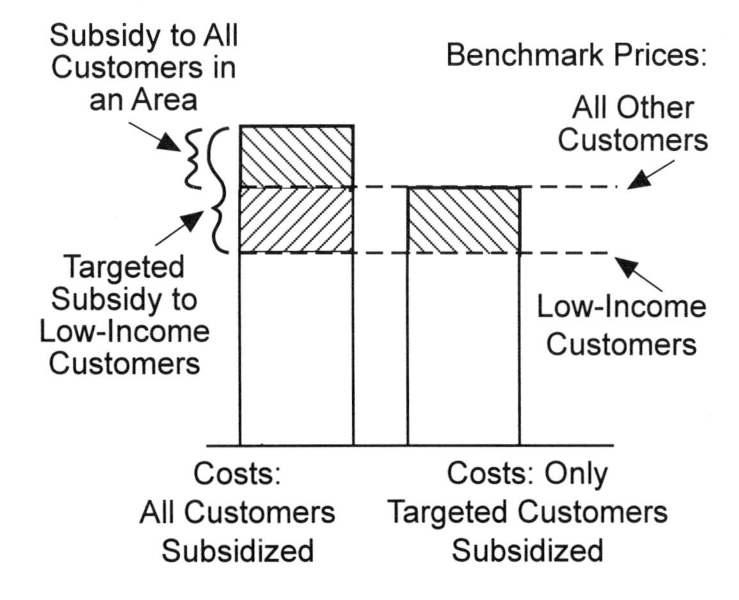

FIG. 8.10. What is one alternative for determining the amount of subsidy? The benchmark subsidy method. From Weinhaus, Monroe et al., 1994, *Universal service toolkit, part 2: Beyond cost allocations: Benchmark subsidy method*. Copyright 1994 by Carol Weinhaus and the Telecommunications Industries Analysis Project. Reprinted with permission.

are needed for at least two reasons: It is politically unacceptable to have major shocks to companies and customers; on the first day of the transition to a new structure (phase in), the world should look exactly the same as on the last day of the old structure (phase out). It is difficult to make a major change when the outcome is unknown. Often people will choose not to act if the result is uncertain, especially if they believe they may be adversely affected.

The enormous amounts of money involved, major political fallout, and other consequences serve as barriers to immediate sweeping changes to the current system. However, if the markets continue to move rapidly due to the forces of technology and competition, major dislocations may occur if no short-term actions are taken. Figure 8.11 depicts a number of short-term solutions, which, when linked with one another, may help lead to major revisions of the current system. Each solution shows a transition from the current mechanism to a new one. For each short-term solution, the first day of the transition must look the same as the last day of the current system.

Who Pays the Subsidy?

During the transition, current subsidies are phased out and eventually replaced by support payments from the new funding source. Prices that tra-

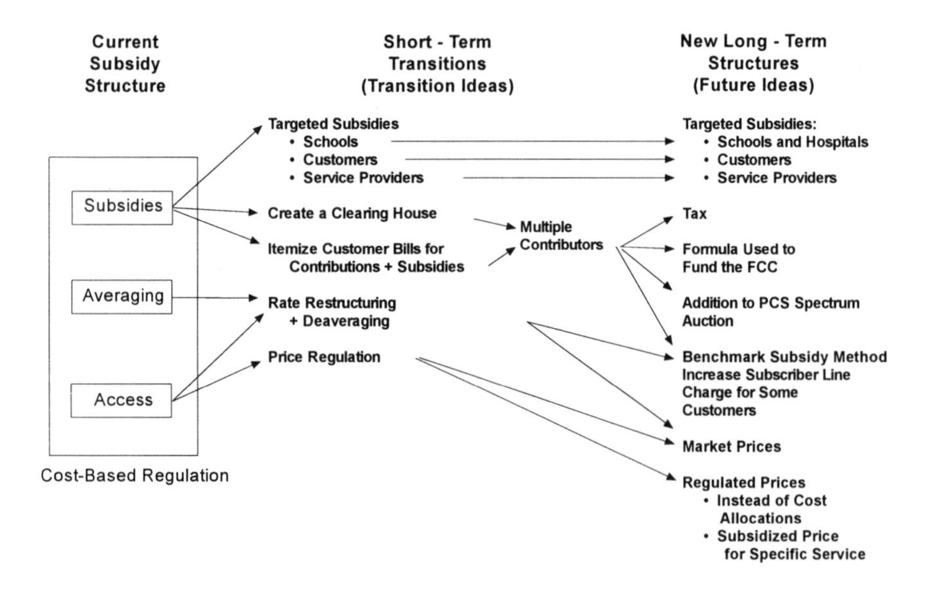

FIG. 8.11. How do you transition from the current environment to the future? Flow from current subsidy structure with transitions leading to new long-term structures. From Weinhaus, Monroe et al., 1994, *Universal service toolkit, part 2: Beyond cost allocations: Benchmark subsidy method*. Copyright 1994 by Carol Weinhaus and the Telecommunications Industries Analysis Project. Reprinted with permission.

ditionally fund subsidies decrease at the same rate that new subsidy funding becomes available.

One method of revising who pays the subsidies eliminates the current patchwork in which some companies' customers pay and other companies' do not. The new mechanism uses some form of tax (e.g., excise tax) on all communications companies (equipment suppliers as well as service providers). The example depicted in Fig. 8.12 shows the relative contributions different industry groups would make to required subsidies if contributions were collected through a fixed and common levy on the value of shipments or revenues.

How Do You Phase in the New Subsidy?

Figure 8.13 depicts a transition from a current subsidy mechanism to a new one. In Fig. 8.13, the subsidy is phased out and replaced by dollars from a

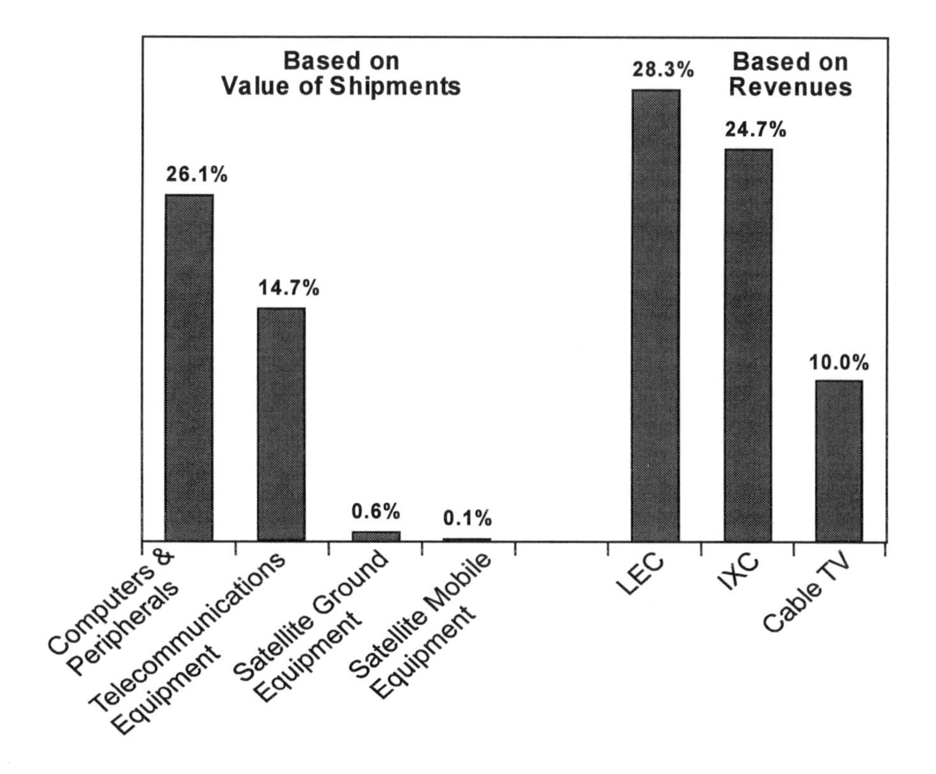

FIG. 8.12. Who pays the subsidy? One example: allocation of contributions among 1992 communications equipment suppliers and service providers. From Weinhaus, Monroe et al., 1994, *Universal service toolkit, part 2: Beyond cost allocations: Benchmark subsidy method*. Copyright 1994 by Carol Weinhaus and the Telecommunications Industries Analysis Project. Reprinted with permission.

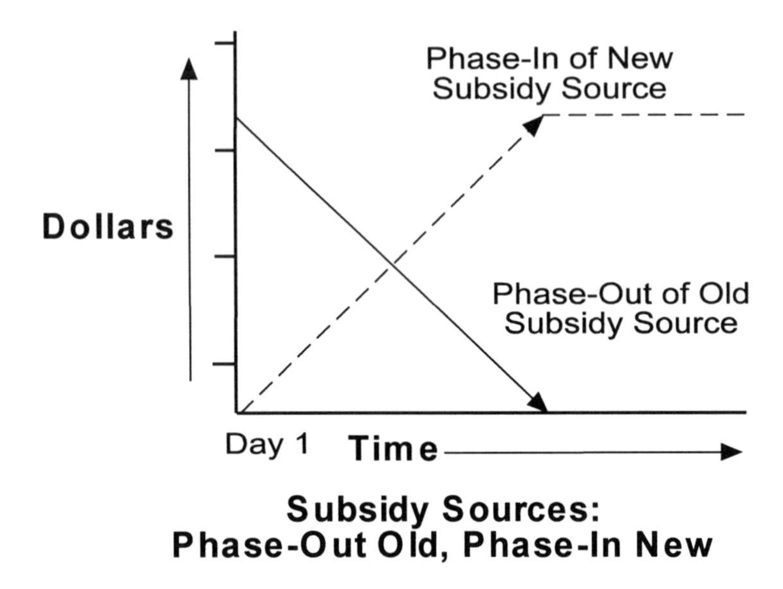

FIG. 8.13. How do you phase in the new subsidy? Phase-out old; phase-in new. From Weinhaus, Monroe et al., 1994, *Universal service toolkit, part 2: Beyond cost allocations: Benchmark subsidy method*. Copyright 1994 by Carol Weinhaus and the Telecommunications Industries Analysis Project. Reprinted with permission.

new source. The subsidy amount on the first day of the phase in is the same as it is on the last day before the transition starts.

On the first day everything looks the same (the straight line with the downward arrow). The phase in of the new mechanism is gradual (the dashed line with the upward arrow). Although the phase in amount is the same as the old subsidy in the figure, in reality the new amount may be more than or less than the old subsidy.

Who Decides What in Communications Policy?

Figure 8.14 indicates why change is so difficult to make—government agencies and courts overlap. Multiple agencies and courts repeat decisions, and sometimes the decisions do not fit with one another. Also, people are able to play one process against another. As a result, it is easier to make small policy changes than to make significant changes that truly alter the political and business landscape. Significant changes are possible only when the most powerful stakeholders agree with one another.

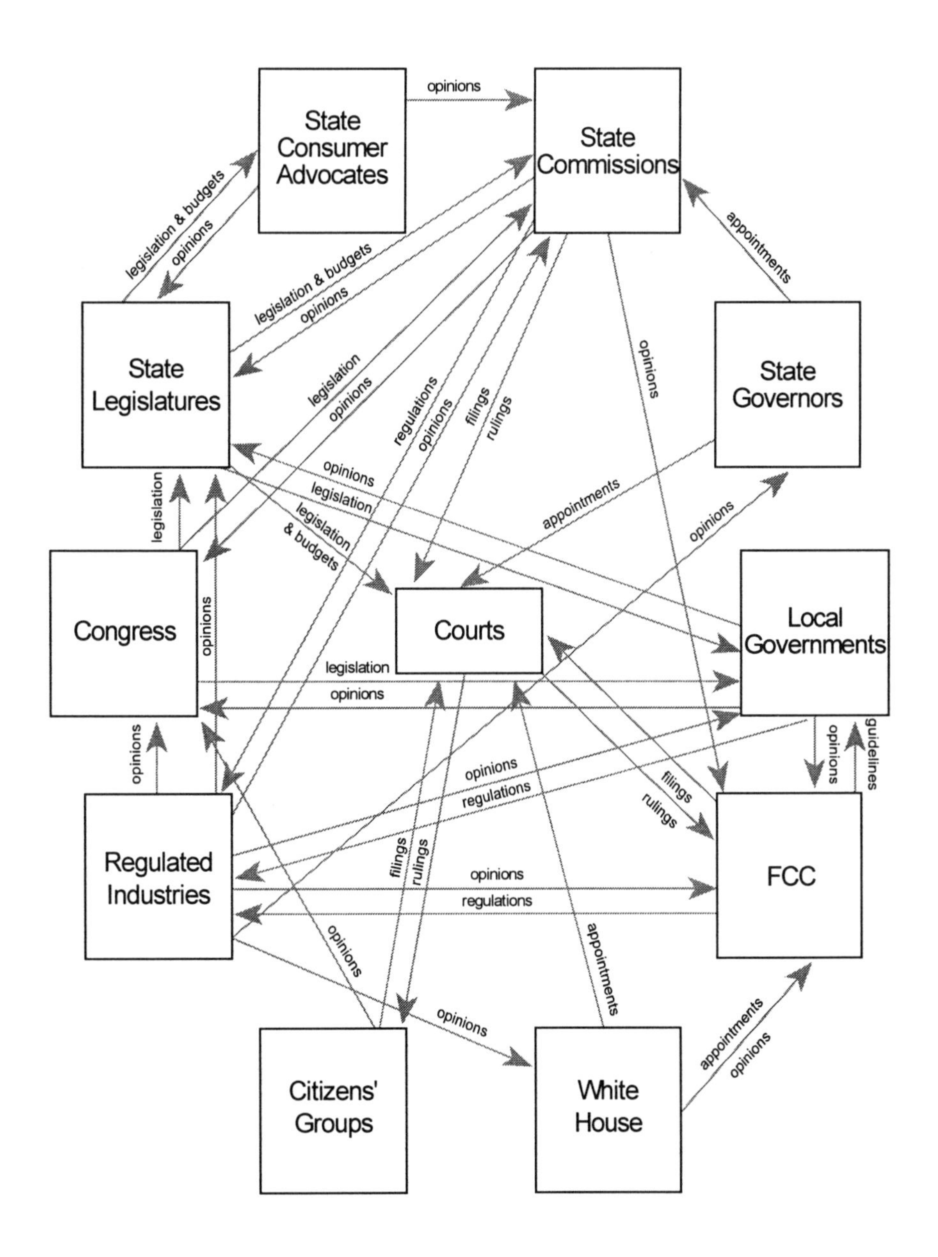

FIG. 8.14. Who decides what in communications policy? It's Hard to Make a Change. From Weinhaus, Pitts, McMillin et al., 1994, *Abort, retry, fail? The need for new communications policies*. Copyright 1994 by Carol Weinhaus and the Telecommunications Industries Analysis Project. Reprinted with permission.

IMPLICATIONS OF CHANGING THE UNIVERSAL SERVICE DEFINITION TO INCLUDE BROADBAND?

What Is the Cost of Expanding the Universal Service Definition to Include Broadband?

Today there is increasing pressure to expand the definition of universal service to include broadband services to all Americans. This is due in part to two major structural shifts that are occurring simultaneously: the shift in market structure from monopoly to competition and the shift in technology from a single provider to multiservice providers. If technologies were not changing, then policymakers would likely examine subsidies in terms of providers, but would not change what is offered. If markets were not changing, then policymakers would wait until a technology was widely accepted before redefining universal service. However, both markets and technology are changing at the same time. The result is confusion over what can be mandated and what should be market-driven.

Figure 8.15 shows the cost of providing a broadband infrastructure and broadband services for nationwide, urban (metropolitan) and rural mar-

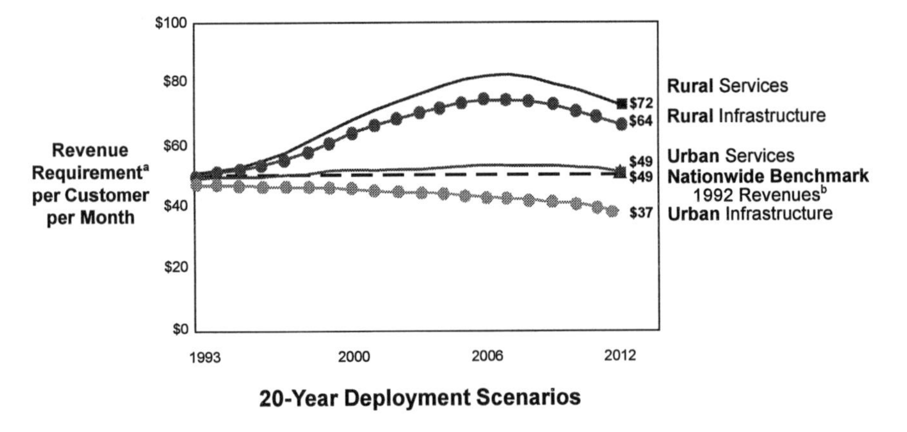

FIG. 8.15. What is the cost of expanding the universal service definition to include broadband? Assumes a 2.35%, annual growth in nationwide access lines. For the broadband services curve, each year 5% of broadband-capable access lines also become equipped for broadband services. The nationwide benchmark of the $48.98 is a calculated revenue requirement for Tier 1 LECs divided by the number of access lines for Tier 1 LECs. Costs are for all access lines (broadband-capable plus nonconverted. From Weinhaus, Pitts, McMillin et al., 1994, *Abort, retry, fail? The need for new communications policies*. Copyright 1994 by Carol Weinhaus and the Industries Analysis Project. Reprinted with permission.

kets. The Nationwide Benchmark line in the figure indicates 1992 local exchange carrier (LEC) operating revenues (including basic service, state toll, and access) per customer per month. It provides a yardstick to give an idea of the magnitude of the costs of deploying broadband.

Figure 8.15 indicates that the costs of providing broadband services to rural areas are significantly higher than in urban areas. One approach to expanding the definition of universal service is the following: When a service becomes required to conduct daily business or personal activities, the service should be included in the universal service definition. The definition of basic service expands only when a technology or service has been widely accepted. Touch-tone is an example of a service that in many areas has been included in the definition of basic service. This approach avoids the issue of some customers paying for services that they do not want. It also allows those customers who do want the new services right away to pay for them and take the risk of picking the wrong technology.

What Happens if You Mandate a Service Before It Is Widely Available?

Figure 8.16 shows the impact of accelerating the deployment of broadband services (networks, training, equipment, and software) to a subset of the nation. Figure 8.16 also shows the incremental investment per student per year needed to provide broadband services over 5 years and over 20 years to the nation's public schools. Both deployment scenarios assume universal

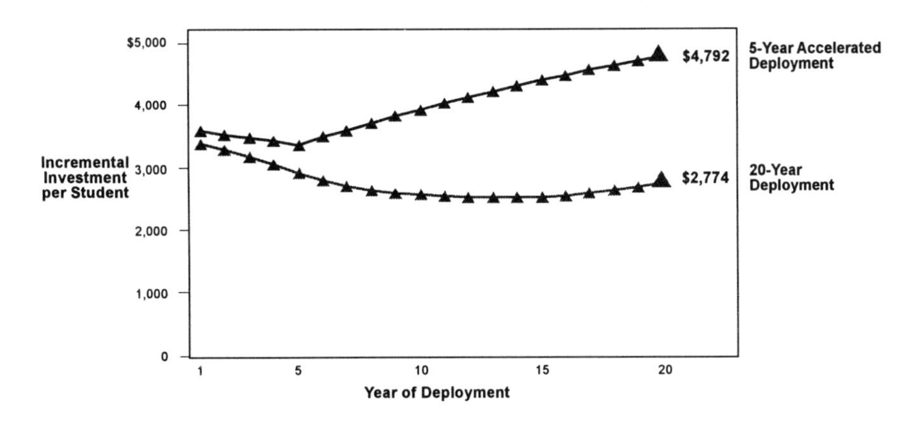

FIG. 8.16. What happens if broadband deployment is mandated? Incremental investment per student per year for universal access to broadband services: public schools, kindergarten through twelfth grade. From Weinhaus, Pitts, et al., 1995, *Schools in cyberspace: The cost of providing broadband services to public schools*. Copyright 1995 by Carol Weinhaus and the Telecommunications Industries Analysis Project. Reprinted with permission.

access (i.e., each student and teacher has a computer). Similar patterns exist in teacher only access and team of students access scenarios.[4] However, in these cases the costs are significantly lower.

In this example, the result of mandating a definition of universal service produces results that are not erased with the passage of time.

SUMMARY

Regardless of your view of subsidies, something has to be done. The current system is crumbling and cannot be sustained under the pressures of new technology and competition. Most people recognize the need for nationwide minimum communications service requirements. These include some voice calling; maintenance and installation; and access to emergency services, operator services, directory assistance, and other networks.

In the future, customers may see simple packages of communications services that offer a step beyond the separate telephone, computer, information and entertainment services of today. Some of these packages will offer services that are not currently combined or available. Customers will pick the package that best fits their lifestyle. This view of the future is called *The Information Studio.*[5] Regardless of what form the future takes, one thing is certain: in the future, customers will have more choices and competition will be prevalent. The current universal service policies, developed in a monopoly environment, will need to be adapted to fit the competitive environment of the future.

ACKNOWLEDGMENTS

The views expressed in this chapter are those of TIAP. The information in this chapter provides general public information and does not constitute or foretell the official position of any of the parties who contributed to this research. The opinions expressed in this chapter do not necessarily reflect the views of the FCC or of any other agency or institution.

TIAP is associated with the Public Utility Research Center at the University of Florida College of Business Administration.

[4]For more details on this figure, see Weinhaus, Pitts, et al. (1995).

[5]For a complete description of this view of the future, see Weinhaus, Monroe, et al. (1995).

APPENDIX A

Background for Rural Deaveraging Model

Sources for Fig. 8.9: Common Carrier Bureau (1992), Taylor (1980).

Assumptions and Caveats for Fig. 8.9:

There would need to be a transition mechanism.

The only objective addressed is universal service. Income transfers and costs of doing business are not considered.

Figure 8.9 shows a nationwide average; it does not show extremes. Customers in areas that have higher than average costs would require support that exceeds this average. Customers in lower cost areas would likely require less support than the average, or perhaps no support.

The $0.7 billion represents an upper bound for the following reasons:

The number of customers assumed to receive support is an upper bound. This assumption is based on the highest customer response to changes in local service prices found in Taylor's survey of telephone demand studies. Using Taylor's data, the assumption is that a 10% increase in price results in a 2% decrease in purchases. The lowest customer response shown in Taylor's survey was one fourth of this estimate—a 0.5% decrease in purchases in response to a 10% increase in price.

Financial support for both local and toll were included. It is possible that both services would not need to be supported because customers can limit their toll bills by placing fewer calls.

All customers receiving financial support were assumed to need support at the current level. Some would probably need less support.

Rural and urban customers have similar demand curves for local telephone service.

Telephone penetration for rural households is the same as the national average.

A mechanism can be designed to target customers that need support—similar to lifeline and linkup programs. (For a discussion of these calculations and sources, see Weinhaus et al., 1993).

APPENDIX B

1995 Participants in the Telecommunications Industries Analysis Project

State Regulators	*NARUC representatives from:*
	Illinois Commerce Commission
	Iowa Utilities Board
	Massachusetts Department of Public Utilities
	New York Public Service Commission
	Ohio Public Utilities Commission
	Washington Utilities and Transportation Commission
Regional Holding Companies	Ameritech
	Bell Atlantic
	BellSouth
	NYNEX
	Pacific Telesis
	SBS Communications Inc.
	US WEST
Independents	Anchorage Telephone Utility
	GTE
	Sprint Local Telecom Division
Interexchange Carriers	AT&T
	Sprint
Cellular and Wireless Carriers	Sprint Cellular
Foreign Domestics	InfoCom Research, Inc.
	NTT America
Local, National, and International Services	BT
Telecommunications Equipment Manufacturers	Nortel
Sponsors	Corporation for Public Broadcasting
Assisting with Public Data	Bellcore
	Federal Communications Commission
	National Exchange Carrier Associations

Background on the Telecommunications Industries Analysis Project

The Telecommunications Industries Analysis Project (TIAP), a 7-year-old research consortium, conducts and reports impartial research in the areas where network planning, business financials, public policy (regulation and legislation) intersect. The participants actively work together to develop new options for telecommunications policies to meet the needs of consumers, governments, and companies in a changing, competitive environment. Participants include regulators, domestic and foreign telecommunications companies, materials and equipment manufacturers, and other communications-based organizations.

The purpose of the project is to produce research and analysis that will assist policy makers in making informed decisions.

TIAP incorporates the following features:

- *Neutral setting.* The Project provides a neutral setting, free of partiality, thereby ensuring objective and independent research.
- *Multiple viewpoints.* Participants play an active role in the research and analysis, represent their own interests, and understand and assist in developing others' perspectives.
- *Analysis and results of alternatives.* The Project provides research data, tools, and models for critical decision making.
- *Public distribution of research.* Data used by this Project are publicly available. Research products become public domain information.

REFERENCES

Common Carrier Bureau. (1992). *Percent of households with telephones.* Washington, DC: Federal Communications Commission.

The Conference Board. (1995). *How we spend our money.* Washington, DC: Consumer Research Center. (Based on a survey by the U.S. Department of Labor, Bureau of Labor Statistics.)

Mitchell, B. M. (1990). *Incremental costs of telephone access and local use,* Santa Monica, CA: The RAND Corporation.

Mueller, M. (1993) Universal service in telephone history: A reconstruction. Telecommunications Policy, 17, 352–369.

Taylor, L. D. (1980). *Telecommunications demand: A survey and critique.* Cambridge, MA: Ballinger Publishing Company.

Weinhaus, C., Makeef, S., Copeland, P., Albright, H., Jamison, M., Bosley, J., Monroe, T., Vasington, P., Harris, D., Berg, S., Lock, B., Pitts, T., Sims, G., Hedemark, F., Monfils, J., Sichter, J., Dunbar, J., Martin, P., Little, L., Garbanati, L., Clark, A., Imafuku, H., Nishioka, Y., & Calaway, G. (1995). *Loop dreams: The price of connections for local service competition.* Boston: Telecommunications Industries Analysis Project.

Weinhaus, C., Makeef, S., Copeland, P., Calaway, G., Jamison, M., Hedemark, F., Harris, D., Monroe, T., Ralston, L., Bond, D., Inman, S., Albright, H., Dunbar, J., Garbanati, L., Sims, G., Adams, S., Monfils J., & Little L. (1993). *What is the price of universal service? Impact of deaveraging nationwide urban/rural rates*. Boston: Telecommunications Industries Analysis Project.

Weinhaus, C., Makeef, S., Jamison, M., Albright, H., Garnabati, L., Calaway, G., Harris, D., O'Brien, M., Hedemark, F., Copeland, P., Sims, G., Adams, S., Sichter, J., Inman, S., Harrell, B., Thönes, R., & Connors, K. (1992). *Who pays whom: Cash flow for some support mechanisms and potential modeling of alternative telecommunications policies*. Boston: Alternative Costing Methods Project.

Weinhaus, C., Monroe, T., Jamison, M., Berg, S., Harris, D., Martin, P., Copeland, P., Little, L., Makeef, S., Pitts, T., Hedemark, F., Bosley, J., Albright, H., Dunbar, J., Garbanati, L., Sims, G., Charlton, D., & Monfils, J. (1994). *Universal Service Toolkit, Part 2: Beyond Cost Allocations: Benchmark Subsidy Method*. Boston: Telecommunications Industries Analysis Project.

Weinhaus, C., Monroe, T., Jamison, M., Harris, D., Albright, H., McCarthy-Ward, P., Monfils, J., Hedemark, F., Dupont, D., Sims, G., Charlton, D., Berg, S., Vasington, P., Makeef, S., Pitts, T., Clark, A., Copeland, P., Martin, P., Garbanati, L., & Bosley, J. (1995). *The information studio*. Boston: Telecommunications Industries Analysis Project.

Weinhaus, C., Pitts, T., Garbanati, L., Bosley, J., Jamison, M., Albright, H., Berg, S., Charlton, D., Clark, A., Makeef, S., Monfils, J., Monroe, T., Vasington, P., Harris, D., Sims, G., McCarthy-Ward, P., Martin, P., Copeland, P., Hedemark, F., Dupont, D., Little, L., Imafuku, H., Nishioka, Y., Lock, B., & Rizzo, C. (1995). *Schools in cyberspace: The cost of providing broadband services to public schools*. Boston: Telecommunications Industries Analysis Project.

Weinhaus, C., Pitts, T., Jamison, M., Harris, D., Berg, S., Dunbar, J., Charlton, D., Copeland, P., Albright, H., Bosley, J., Garbanati, L., Monfils, J., Martin, P., Makeef, S., Monroe, T., Clark, A., & Sims, G. (1994). *Apples and oranges: Differences between various subsidy studies*. Boston: Telecommunications Industries Analysis Project.

Weinhaus, C., Pitts, T., McMillin, R., Jamison, M., Albright, H., Harris, D., Monroe, T., Dunbar, J., Garbanati, L., Berg, S., Copeland, P., Charlton, D., Makeef, S., Little, L., Hedemark, F., Martin, P., Sims, G., Monfils, J., & Cowles, R. (1994). *Abort, retry, fail? The need for new communications policies*. Boston: Telecommunications Industries Analysis Project

9

Recovering Access Costs: The Debate

David Gabel
Queens College

The Conference Report for the Telecommunications Act of 1996 states that the overall purpose of the law is "to provide for a pro-competitive, deregulatory national policy framework designed to accelerate rapidly private sector deployment of advanced telecommunications and information technologies and services to all Americans by opening all telecommunications markets to competition" (*Conference Report for the Telecommunications Act of 1996*).

The federal government has clearly established a goal of promoting competition. This chapter addresses ways in which the cost of customer access would be recovered in a competitive market. I focus on customer access because it is the largest cost center associated with providing universal service. Many utility economists have attributed the need for a universal service fund to the regulated price of residential service and/or customer access. The local exchange carriers (LECs) have told commissions repeatedly that competitive market pressures require that this cost be recovered through the price of exchange service.

Proponents of increased charges for local service assert that the cost of the loop should not be recovered through usage charges. They claim that it is economically inefficient to impose a tax on usage-related services in order to help cover the cost of providing customers with access to the network. Currently, the common carrier line charge recovers a portion of the loop's cost through the price of toll services. The inflated toll price allows customers to obtain access to the network at a reduced rate (Kahn & Shew, 1987; Kasserman & Mayo, 1994). Kahn (1984) characterized these price levels as inefficient because they discourage use of long-distance calling and encourage overconsumption of network access. Kahn does recognize that there are two economic reasons why the price of exchange service may not cover the total cost of access—society may want to provide aid to those customers who can least afford telephone service, and because the value of network subscription decreases when one customer disconnects service, the externality of telephone service should be reflected in the price for marginal customers.

Some of the economic arguments for higher customer access charges are perverse. For example, Kasserman and Mayo make the illogical assertion that the fixed costs of the firm should be recovered exclusively through a fixed customer charge. When they write "fixed costs, by definition, bear no relationship to the volume of usage" (Kasserman & Mayo, 1994, p. 124), the logical correlative statement would be neither are fixed costs attributable to providing customer access. Rather than making this statement, they assert that all fixed costs should be recovered through the price of customer access.

Economists adopt efficiency as one standard for judging the efficacy of a pricing structure. They maintain that society's welfare is enhanced when economic efficiency is raised. Kahn believes that efficient prices are the inevitable outcome of competition. It is this conviction that motivates his argument that regardless of how politically unpalatable, regulators should recover the nontraffic-sensitive costs exclusively through the exchange service rates. Furthermore, as the telecommunications market becomes increasingly competitive, there will be an increased need to raise the price of exchange service. "The one thing that is certain is that the new regime of competition, on the one hand, and the perpetuation of the old regime of inefficient pricing, on the other, are fundamentally incompatible; one or the other is going to have to give" (Kahn, 1984, pp. 150–151). Kasserman and Mayo, among others, support Kahn's proposition that competition is forcing the nation to move to a more cost-based pricing structure. They assert that society's welfare will be enhanced by this transition but present little or no meaningful evidence to support their proposition that, in competitive markets, all customer access costs are recovered through a fixed customer charge (Kahn, 1984; Kahn & Shew, 1987; Kasserman & Mayo, 1994).

Kahn's argument concerning the superiority of cost-based prices rests on the assertion that society's welfare is maximized if customers are charged a price that reflects the marginal cost of production. This is a fundamental tenet of economics and the logic is quite compelling. If customers are charged more for a product, such as long distance, than the cost society incurs in providing the product, the equilibrium level of output will be less than the amount that maximizes society's welfare. Consumers will purchase long-distance service up to the point that the additional benefit equals the additional cost. If the price for toll service is set at a level that exceeds the costs that society must incur when producing the product, consumers will stop purchasing the product at a point where the value to a consumer is less than the cost society incurs in making the product available. Society's welfare would be maximized if the level of output is at the point where the value of the last call is equal to the cost society incurs in providing the call.

Kahn and other economists have argued for years that toll consumption is suppressed because the price of long-distance service is inflated by the common line charge. The price reflects a portion of the nontraffic-sensitive cost of the loop, a cost that is not affected by the volume of toll calls. The price of the marginal toll call should not recover a portion of the cost of the loop because the cost of the network connection is independent of the volumes of calls.

COMPETITION IS THE BENCHMARK

Kasserman and Mayo (1994), as well as Kahn, argue that competition will inevitably eliminate this inefficiency: "Competition inexorably drives prices to marginal costs" (p. 137). These economists contend that the commissioners should quickly raise the price of exchange service and lower the price of toll calls. These actions, they assert, would be consistent with competitive market behavior. Despite Kahn's appearance in numerous proceedings, few states have adopted their suggestions. Reflecting on the activities of regulatory commissions, Kasserman and Mayo asked recently:

> Why have regulators continued to cling to this complex web of cross-subsidies (to the extent mounting competition and bypass permit) despite the desirable attributes of efficient pricing, the undesirable attributes of the existing cross-subsidies, and the challenge to rationalize telecommunications pricing Kahn issued a decade ago? ... If regulators do not respond to the arguments we present, it may very well be because we are presenting the wrong arguments or presenting the right arguments wrongly. (pp. 120, 142).

These authors maintain that if regulators were working in the public interest, they would adopt the "efficient" pricing standards proposed by Kahn. According to them, raising the price of exchange service would eliminate the need to recover a portion of the loops through toll rates. Kasserman and Mayo believe that it is political pressures that cause the commissioners to ignore their economic arguments and to opt instead to sustain inefficient prices. The commissioners, these authors assert, fear the political backlash that would occur if the price of local service was increased (Kasserman & Mayo, 1994).

Kasserman and Mayo are optimistic that competitive market forces will be stronger than the nefarious influence of politics. That "[t]he increasing intensity of competitive market forces in this industry will eventually necessitate the recommended pricing structure regardless of regulatory action or inaction because competition inevitably drives prices to marginal costs" (Kasserman & Mayo, 1994, p. 121).[1] Kahn and Shew (1987) offered a

---

[1]A similar lack of understanding of how pricing evolved in the competitive, unregulated telecommunications market has been demonstrated in a working paper by Bruce Egan and Steven Wildman, *Funding the Public Telecommunications Infrastructure*: "This system, rooted more in political compromise than economic logic, was sustainable when there was a single monopoly network. However, competition is the natural enemy of cross-subsidy and the emergence and continuing growth of competitive suppliers makes it doubtful that the current mechanisms for managing these transfers can be maintained much longer" (1994, January, p. 4).

similar sentiment; that as rivalry increases, all else being equal, the greater the likelihood that a firm will charge an access fee to recover customer access costs.

Kasserman and Mayo conclude that the failure to adopt higher access fees is due to political influence. I argue that their alternative hypothesis—that utility economists do not offer an explanation that is consistent with how markets work—is a better way to explain their regulatory defeats. The regulatory record suggests that some commissions, Florida being a notable example, start with a view of how a fully and effectively competitive market works, then they devise policies that replicate those conditions.

COMPETITIVE MARKET BEHAVIOR

In a proceeding before the Florida Public Service Commission, interexchange carriers (IXCs) advocated that they not be made to pay for use of the local loop because the facility is nontraffic sensitive. The Commission rejected this proposition, noting that it would be "contrary to common business practices which is to charge customers for use of fixed cost facilities in the price for goods and services. It is appropriate that each service provide some contribution toward the fixed costs common to those services" (Investigation into Nontraffic Sensitive Cost Recovery, 1987). The Commission's position is just the type of regulatory decision with which Mayo, Kasserman, and Kahn disagree. They feel that it fails to adapt to the imperatives of competitive market forces. In the following sections, I provide examples to illustrate that the Commission was correct. In competitive markets, customer access line costs are not recovered entirely through the price of access (*In the Matter of the Bell Atlantic Telephone Companies*, 1995).[2] Rather, the competitive standard requires that the cost of access should be recovered from all the products that benefit from the facility. I provide these examples for three reasons. First, the 1996 Telecommunications Act makes it clear that it is the nation's policy to promote competitive outcomes. Recovery of customer access costs exclusively through customer access charges is contrary to the law [Section 254(K)]. Second, the emulation of competitive markets is a widely recognized objective of regulation. I believe these examples illustrate that the elimination of the common carrier line charge would violate this standard. Finally, a number of economists, such as Kasserman and Mayo, have claimed that value-of-service pricing is

[2]It should be noted, however, that although Dr. Taylor supported the principle that "the common cost of the network platform should be recovered from all services that use the platform in this affidavit, in an affidavit filed in the universal service docket with Kenneth Gordon, he takes the opposite position with regard to the sharing of loop costs, although the loop is used by all switched services (Gordon, Kenneth & Taylor, William E., Comments on Universal Service, p. 8 [April 12, 1996]; attachment to *Comments: In the Matter of Federal-State Joint Board on Universal Service*, Bell South Corporation CC Docket No. 96–45 [April 12, 1996]).

not sustainable in competitive markets (Kasserman & Mayo, 1994). Both the theoretical and empirical literature provide many counterexamples that illustrate the fallacy of their assertion. Casual empiricism suggests that value-of-service pricing is more the rule than the exception.

The next section looks at how competitive markets work. First, I present a brief overview of how several modern markets operate in a competitive environment. I then provide a detailed look at how the telephone industry operated historically under conditions of rivalry.

SEGMENTED PRICING

Noticeably absent from the writings of Kahn, and other like-minded economists, is any reference to the theoretical literature, which suggests that under conditions of rivalry, the cost of access is recovered entirely through a customer access charge. Indeed, the theoretical literature shows just the opposite. Scotchmer (1985) showed how as the number of suppliers in a market increase (i.e., the degree of competition increases), customer access fees converge to zero.

What Kahn, Kasserman, and Mayo seem to have overlooked in their analyses is that people do not make purchases by evaluating the products alone but by evaluating the entire purchase opportunity. This makes a firm's pricing decisions as much a function of strategic positioning and marketing as a function of cost recovery. "Moreover, the cost of servicing different buyer segments, and the intensity of the competition to serve them, also varies greatly for the same product. Consequently, effective pricing often requires a strategy of segmented pricing."[3] This type of strategy takes many forms, some of which are: segmentation by peak-load pricing, such as is seen in the electric utility and telecommunications industries; and segmentation by product design, which can be found in the transportation, software, and retail industries. The variety of strategic pricing options that are used by firms under conditions of competition, of which the previous are just a small sample, imply the existence of a much more dynamic and fluid pricing environment than the one envisioned by the utility economists.

In the following paragraphs, I provide brief sketches of firm behavior as empirical support for the proposition that the cost of access is not recovered exclusively through an end-user charge in competitive markets.

---

[3]Southern New England recently expressed a similar view to the Connecticut Department of Public Utilities: "competitors will look at the total basket of services a customer buys, not any single service, in making a decision to market to that customer. Put differently, it is the profitability of individual customers, not individual services, that is attractive to competition" (*Response of Southern New England Telephone*, TE052 Supplemental Response, [October 13, 1995]; application of the Southern New England Telephone Company for Financial Review and Proposed Framework for Alternative Regulation.)

TRANSPORTATION AND VALUE-OF-SERVICE PRICING

Telecommunications is hardly the first industry in which economists have taken the position that value-of-service pricing is incompatible with competition. For a number of years, economists ascribed value-of-service pricing in the transport industry to regulation as well. They argued that if the industry were deregulated, competitive forces would lead to a cessation of value-of-service pricing. In the past decade, the deregulation of the railroad and truck carriers has provided a laboratory for testing the hypothesis that value-of-service pricing would not be retained in a competitive, deregulated market. Truck transport has been characterized as competitive because of "free entry, many buyers and sellers, and considerable flow of information" (Beilock, 1985, p. 94). The empirical work suggests that value-of-service pricing continues to be practiced in the competitive, deregulated transportation markets, as well as those transportation markets that were never regulated (Beilock, 1985; Talley, 1989).

The persistence or growth of value-of-service pricing has also been observed in another recently deregulated segment of the transportation industry—airlines. In their study of airline pricing practices, Borenstein and Rose (1994) found that the amount of price discrimination actually increases with the degree of competition.[4] When challengers are present, suppliers use pricing schemes that are designed to lock-in customers. By offering tourists large discounts, an airline can attract these consumers away from other suppliers, while maintaining large markups for business customers. When there is less rivalry, an airline has a reduced incentive to pursue business. Because there is no other game in town, suppliers are less aggressive about attracting the price-sensitive consumer.

Vietor (1994) summarized the impact of deregulation in six industries. He points out that to the surprise of many proponents of deregulation, pricing mechanisms became more complex once government controls were reduced. Rather than moving to cost-based pricing, as many economists had predicted, many of the markets exhibited an increased level of price discrimination because firms used pricing to segment customers and establish customer loyalty.

CREDIT CARD COMPANIES AND ANNUAL FEES

Many credit card companies, despite the option value of the service, do not charge an annual fee to certain customers, although setup costs plus

[4]Price discrimination occurs when a firm sells similar goods at rates that are in a different ratio to marginal costs. (Stigler, 1987).

monthly billing surely lead to access costs (Ausubel, 1991).[5] During the 1970s, banks wanted to establish annual fees for credit cards, but they refrained from doing so because of competitive market pressures. Banks that tried to impose these charges experienced a large decline in their number of cardholders. Not until 1980, when the Carter Administration imposed new rules on reserve requirements, were the banks able to act successfully in concert and establish membership fees (Mandell, 1990). Nevertheless, today's customers can obtain credit cards for a zero access fee precisely because entry into the credit card industry is relatively unimpeded. This example is actually closer to home than it first appears: AT&T's entry strategy in the credit card market was to offer a lifetime waiver of the annual fee.[6]

American Express attempted to continue to serve customers exclusively through cards that required an annual fee. Their failure to respond quickly to customers' clear preference for cards without an annual fee caused them to lose substantial market share. In 1996, *The New York Times* reported that "Two years ago, the company shifted its focus from its pay-as-you-go charge cards, with their lucrative fees, and belatedly showered the public with what clients really wanted: interest-bearing cards in a dozen flavors, many of them free" ("The Card," 1996, p. D1).

The variety of pricing structures in the credit card industry is consistent with my belief that if the exchange market were competitive, competing carriers would offer customers a menu of pricing plans. Although some customers would choose to pay a high membership fee in exchange for a low usage fee, others would prefer the reverse. The evidence for price discriminating behavior and market segmentation is further highlighted by the case of Gillette.

GILLETTE AND THE POWER OF PRODUCT SEGMENTATION

A firm that has learned the art of strategic pricing extraordinarily well is Gillette. Gillette has chosen to focus on a "shaving systems" approach in order to take full advantage of

> the principle of complementary products under which the relative prices of products can be exploited because they must be used together. The razor, a quite substantial product, is sold at low price to get it into the consumer's hands. This facilitates the sales of profitable, replacement blades which fit only the systems for which they have been designed. (Thomsen, 1987, p. 44)

---

[5]Ausubel describes how consumers seem to choose consistently cards with a low annual fee (analogous to the access charge), even when the package entails a heftier annual borrowing rate than a package with a higher fee.

[6]And a Big Loss at AT&T. (1996, September 16). *Business Week*, p. 54. AT&T's marketing strategy has been quite successful. The card was introduced in March of 1990. As of September 1996, AT&T's card was the third-largest credit card in terms of accounts.

Gillette has carried this principle further by using its market dominance to influence point-of-sale (POS) displays so that all of Gillette's shaving and shaving-related products are grouped together to enhance the image of Gillette as a quality purveyor of men's shaving products.

A Gillette analyst has pointed out that another component of the company's strategy has been to continually add features to the basic razors, and hence make more profit per blade as consumers buy the features. This started with the Trac II twin-blade system, and continued with the pivot head first on the Atra, and then later on the Good News disposable. Following this introduction was the addition of a lubricating strip on the blade that would release a lubricant when wet. This feature was first put on the Atra Plus, and later added to the Good News Plus.

What Gillette has been doing effectively is hooking the consumer with a low-priced razor and blade, and then having the consumer buy upscale a little each time. With a fixed market size, this is almost the only way to increase profits (Thomsen, 1987).

To augment the previous strategies, and to ensure that their brands offer value, Gillette also follows what it calls a market basket approach to pricing. "The company keeps a daily track of a collection of lowly items, including a newspaper, a candy bar, a can of coke, all ranging from 10 cents to a dollar. And then it never raises its prices at a faster rate than the price of this market basket" (Morris, 1996, p. 5). This profit maximizing strategy is driven by providing value to customers, not setting price equal to cost.

NETSCAPE AND THE POWER OF INCREASING RETURNS

Another pricing/marketing strategy that has been highly successful is the one embodied in the idea of increasing returns. The basic idea of this is that once you create a big market share for your product, there will be a strong tendency for that market presence to expand. This is especially true in the case of the software industry, which is highly susceptible to network externalities. The more widely used a product is, the more value people will place on it. This value increases even more if the company that makes the initial product can persuade third-party vendors to create products that complement it (Aley, 1996). In addition, software is a high-fixed cost/low-variable cost product. Research and development is the most expensive component of any software package. The actual production of each shipped unit is relatively small. It involves little or no raw material cost, little or no inventory cost, and with distribution over the Internet, distribution costs can be almost zero (Aley, 1996). It is this set of ideas and circumstances that are behind the seemingly irrational actions of companies, such as Netscape, that distribute software free of charge through the Internet.

The Netscape Navigator made its debut in December 1994 and within 7 months of its release, two thirds of the 9 million browsers used on the WWW were Navigators (Sprout, 1995). This giveaway not only resulted in a massive market dominance for Netscape's Navigator software, it also prompted thousands of third-party vendors to write add-on products that were designed to work with Navigator. Some of these add-on products are available only in the enhanced version of Netscape, which is available for $39 and which also includes customer support (Sprout, 1995).

By giving away its browser software, Netscape created a high-profile awareness of its product and a broad base of users and vendors. The company capitalized on this awareness, and its base of specialty add-on vendors, to leverage its position as a seller of more highly priced corporate server and browser software packages, which are specially configured to meet the growing needs of corporate Internet communication (Sprout, 1995).

Another company which has played this give-it-away-and-thrive game is id Software, the creators of Doom and Quake. The company spent 2 million dollars creating Quake and then distributed an abbreviated, but fully functioning version of the game free on the Internet. To obtain the game in its entirety, with all its bells and whistles, a user has to buy it in a store or via an 800-number. This strategy has "put its games on millions of hard drives around the world—and made $16 million in 1995" (Aley, 1996, p. 1).

As a caveat, it should be noted that software development of this kind involves large fixed costs that are not specific to a customer. This cost structure is much different from a telephone network for which there are access costs specific to customers. However, the interesting point here is that Mayo and Kasserman argue that fixed costs, such as software development costs, should be recovered from a fixed customer charge. Clearly, as the previous examples make explicit, not all software firms follow this strategy and those that do not often do quite well for themselves in this ferociously competitive market. Indeed, the Netscape practice of providing end-user access for free in order to create value for the product has been deemed by *Business Week* "the business model for the new software industry" ("The Software Revolution," 1995, p. 90).

INTERNET SERVICE PROVIDERS AND ACCESS PRICING

The provision of Internet access and service is another highly competitive market, one in which if Kasserman and Mayo are correct, all customer access costs would be recovered through access charges. What is actually observed, however, is something a little bit more complicated, as Table 9.1 makes clear.

In Table 9.1, we can see that approximately half of the Internet service providers (ISPs) do not charge setup fees to prospective customers. This lack of a setup fee is used primarily to lure customers to sign onto a certain

TABLE 9.1

Selected Internet Service Providers* (ISPs)

Total Number of ISPs	*Percent of ISPs Requiring a Setup Fee*	*Range of Setup Fee*		
50	40%	$10 to $50		
Monthly Fee Charges: Percentage of ISPs				
$5 to $10	$10.50 to $15	$15.50 to $20	$20.50 to $30	
8%	12%	44%	36%	
Hours Included in Monthly Fee: Percentage of ISPs				
0–20	25-50	60-100	125-240	unlimited
12%	14%	20%	4%	46%

* Offering Personal PPP or SLIP Accounts accessible with a Local Analog Call from within the greater Boston Area.

network.[7] Furthermore, it is evident that a significant proportion of ISPs offer unlimited access to network subscribers. Offerings of unlimited access by this great a proportion of access providers in an industry whose cost structure is highly usage sensitive would seem to indicate that competition is not driving all ISPs to create usage pricing structures.[8] This is especially surprising given the fact that as the network becomes more congested, the provider will have to lease out more lines/bandwidth, upgrade modem pools, and incur other costs related to the increased usage. Obviously, it would appear that competition in the provision of Internet access is not driving price to cost. Instead, the story that Table 9.1 illustrates is that ISPs are segmenting their markets by offering consumers a variety of variably

[7]Another example of this approach is MCI's decision to offer members of its Friends & Family Program a free electronic mailbox. (1994, November 11). *Wall Street Journal.* p. B3, quoted in: Anderson, Bikson, Law, & Mitchell (1995), p. 15.

[8]Part of the lack of usage based pricing might be due to technological constraints. The most obvious usage charge, a charge per packet, would consume more computer capacity than is needed to transmit the packets in the first place. "Freeloading as a way of Life; The Strange Economics of Cyberspace" (Copyrighted 1995). *Economist.* [http:\\www.economist.com].

priced bundling options from which to choose. In other words, the ISPs are responding to the pressures of competition by pursuing value-of-service pricing strategies similar to those that the transportation industry and airlines followed on deregulation.[9]

PRICING OF ACCESS IN THE MOBILE MARKET

The value of rivalry in introducing change in pricing structures has been richly described in a recent Organization for Economic Co-Operation and Development (OECD) paper: "Mobile Communications: Pricing Strategies and Competition." The OECD (1995) found that when one firm controls the market, this supplier tends to view the market as static; consequently, it does not actively seek new customers or experiment with new pricing plans and products. Although the LECs have seen access traditionally as a burden associated with providing telephone service, mobile suppliers have lowered the price of access because of the drive to obtain market share. Rather than focusing on the handset as a cost center, the entrants have heavily discounted the price of access. They know that to be successful in a competitive market, it is essential to have customers become members of their networks. As has recently been pointed out in *Business Week*, "Cell-phones increasingly are seen by retailers as little more than giveaway commodities, useful for signing customers up for service." ("Motorola Goes For the Hard Cell," 1996, p. 39). The cost of providing free phone sets has hardly been a trivial expense, the "free" hardware typically costs the network operator $150 per handset. The access provider offers the free telephone because it wants to gain subscribers for the network. Then it can earn profits from commissions on air time.

The recent pricing developments in the mobile market raise the issue: if the wireline telecommunications market were competitive, what type of pricing structures would be observed? The experience of the wireless market provides further support for my proposition that rivalry would promote a proliferation of different pricing structures, including those with reduced customer access fees. If the market story offered by Kahn, Mayo, and Kasserman were correct, we would not see mobile companies giving away telephones in order to provide customer access. Instead, we would observe a persistence of the pricing structures that existed during the era when there was less rivalry—customers would have to make a one-time payment for their handsets.[10]

---

[9]"Customers will face a variety of optional price structures and will be encouraged to purchase bundles of services with inducements to stay loyal (term commitments, frequent flyer tie-ins, cash back every quarter, etc.)," (Bikson et al., 1995, p. 15).

[10]The cellular operators recover the cost of providing customer access through usage fees and a fixed monthly fee. The fixed monthly fee typically provides a minimal amount of air time. The fixed monthly fee does provide a means for recovering a portion of the cost of the mobile set.

ACCESS PRICING IN THE INTEREXCHANGE MARKET

During the late 1970s and the early 1980s, some suppliers entered the long-distance market. These other common carriers, such as Western Union, Sprint, and MCI, originally charged their subscribers a fixed monthly access fee. However, due to competitive pressures, they eliminated it. Sprint was the first to drop the charge, in January 1984, for those customers whose bill was greater than 5 dollars per month. According to a December 16th, 1983 *New York Times* article, "the policy change is Sprint's attempt to increase its market share quickly as competition for lucrative long-distance calling increases" (p. D1). Sprint's pricing change illustrates that competition often results in either a negligible or nonexistent fixed fee. Contrary to the predictions of Mayo and Kasserman, this competitive firm did not use a fixed customer fee to collect its customer-specific costs, such as billing, or the fixed costs of the firm's operations.

TELEPHONE PRICING TRENDS UNDER COMPETITION

The archival records of AT&T demonstrate that the claims that value-of-service pricing is antithetical to competitive market behavior is no more true than the claims mentioned earlier about the transportation, airline, or other industries. Prior to 1907, AT&T provided local exchange and toll service, but it was not regulated. In 1894, when its patent monopoly expired, it faced competitors in many markets. This period of rivalry provides experimental data regarding the claims of economists that value-of-service pricing is not sustainable in competitive markets. At the turn of the century, as in the airline industry today, value-of-service pricing or price discrimination increased with the degree of rivalry. The managers of the various telephone companies concluded that in order to maximize their profits, price discrimination should be used to encourage network membership. The archival records clearly show that the promotion of residential service was not a matter of state public policy, but rather was grounded in sound economics. Competition increased the degree of price discrimination, contrary to the prognostications made by Kahn, Kasserman, and Mayo. This lesson should not be lost in the current debate; as long as LECs still have monopoly power, they should not be allowed to impose pricing structures that are not sustainable in competitive markets.

The history illustrates a second point. Whenever discussions of raising residential rates arise, some proponents of higher rates claim that the consumer advocates should have only one objective, to ensure that low-income people have access. Granted, society should be concerned that low-income areas have the smallest percentage of households with telephones. Telephone service is a basic link to the outside world and, as such, should be a given. But there is also a sound economic reason why commissions should prevent high prices for residential customers. Local exchange is basically a monopoly and, therefore, one of the main purposes of regulation is to see that regulated prices emulate competitive market behavior. Setting market-based prices in a monopolistic or oligolistic industry requires some government policy; so clearly this is an issue of regulation, not merely social welfare. The following history illustrates that the advent of rivalry caused Bell to give discounts to residential customers, discounts that increased as the competition heated up. Also, the incumbent finally began to develop rural markets that had been largely ignored until the Independents showed Bell their potential profitability. In short, rivalry spurred Bell to do a better job of developing telephone service. Thus, although low residential rates may benefit low-income communities, the deciding factor for commissions is that relatively low residential exchange rates are the natural outcome of competition.

When Alexander Graham Bell's patents expired in 1893 and 1894, entrants into the industry, known collectively as the Independents, challenged AT&T extensively and intensively. Bell chose to ignore rural markets because it believed that the marginal efficiency of capital was higher in cities ("Wisconsin Telephone News," 1906). In the cities, Bell had largely ignored the residential market (Fischer, 1992).

Rivalry, along with some comparatively minor factors, caused the average revenue per station to fall from $90 in 1894 to $42 in 1907. Concurrently the number of subscribers increased from 270,381 to 3,839,000 (Federal Communications Commission, 1939). During this competitive period, price discrimination increased. Nationally during the monopoly period, AT&T priced business service at a 20% premium relative to residential customers. During the competitive era, the premium increased to 52% (AT&T Annual Report, 1909).

The average revenue per subscriber reflects two simultaneous affects; shifts in the mix of goods purchased and changes in the price of service. This mix can be controlled for by looking at the prices in a particular market. AT&T had a monopoly in St. Louis from 1876 to 1894. From the expiration of Alexander Graham Bell's patents in 1894 until 1910, there were two suppliers in the city. Under conditions of rivalry, the industry experienced unprecedented growth. During the first 5 years of competition, the number of customers almost doubled relative to what had been achieved during Bell's 17 years of monopoly. Table 9.2 illustrates how, in St. Louis, rivalry led to an increase in the premium that business customers paid relative to residential customers. It also multiplied the number of subscribers. During the monopoly era, premium business service cost 20% more than residential service; by 1904 the differential had increased to 125%. In 1994, the average rate for unlimited business and residential service was $44.65 and $19.80, respectively. Hence, the markup in today's monopoly market is slightly less

TABLE 9.2

St. Louis Development and Rates

January 1st	*1894*	*1899*	*1904*	*1909*	*1910*
Development					
Bell stations	3,889	5,121	15,222	41,836	46,312
Independent stations	—	3,200	11,600	19,400	21,400
Bell stations per 100 populations	0.8	0.9	2.3	5.6	6.2
Independent Stations per 100 populations	——	0.6	1.7	2.5	2.8
Rates					
Business					
Bell maximum	$120	$150	$150	$125	$72
Independent maximum		$60	$60	$72	$72
Bell minimum	$100	$39f	$36.50f	$36.50f	$36.50f
Independent minimum		$60	$60	$60	$60
Residence					
Bell maximum	$100	$100	$60	$54	$48
Independent maximum		$36	$48	$48	$48
Bell minimum	$60	$30f	$18.25f	$27.38f	$24f
Independent minimum		$36	$36	$24	$24
Bell average exchange revenue	$96.87	$89.21	$50.38	$39.40	$35.50

Note. f = measured service. From AT&T Corporate Archive, 1909.

than that which existed in St. Louis, 125% (Federal Communications Commission, 1995, pp. 20, 25).

PRICING TELEPHONE SERVICE IN AN UNREGULATED MARKET

The data in Table 9.2 clearly demonstrate that under conditions of rivalry, the extent of price discrimination increased. AT&T's pricing philosophy for residential and rural customers was clearly developed during the 1900 AT&T Presidential Conference, which was attended by the presidents and other officers of various Bell Operating Companies (AT&T Corporate Archive, 1900). The consensus attitude of the officials was that the profitability of a service or an exchange should not be judged by the relationship between the direct cost of providing the product and its price. Instead, the contribution of the enterprise to the firm's profitability should be determined by considering its effect on the total network. E. J. Hall, the President of Southern Bell, AT&T's Buffalo exchange, and a leading official in its long-distance division, lucidly stated the decision rule that should be used to judge the profitability of an endeavor. "[T]he profit need not necessarily be immediately attached to the particular transaction, but that the company itself profits by what is done." Hall emphasized this view by asking his peers a series of questions:

> [D]oes it not come to a question of whether each individual transaction, each thing that we do, should carry a profit with it, or whether the thing that we do it is wise to do if it carries a profit to our general system? That would apply to taking a specially low rate subscriber in an exchange; if it resulted in a larger profit to the exchange; if it resulted in a larger profit to the exchange or ... an entire exchange, if that exchange added to the profits of the company, or strengthened the position of the company, it would be good policy to take it even though in itself it did not directly carry a profit with it. I suppose all of us would agree to that proposition?[11]

No one disagreed with Hall. Rather, throughout the meeting different participants emphasized the need to think of the benefits of a transaction in terms of its network or systemwide effects. These businessmen had a common understanding of how value was created in a network, the larger the

[11]Hall made his comments in support of a statement that had just been made by President Cutler of New York Telephone. Cutler had pointed out that due to competitive pressures, New York Telephone opened some rural exchanges for strategic purposes in order to impede the development of a competitive network: "So that I would not say that I would not open an exchange even, in some cases, where I could not see any profit in the exchange as a whole, that being, of course, a part of the larger system which covered the surrounding country, which was profitable" (AT&T Corporate Archives, 1900, pp.155–158). Economists who proclaim that competitive behavior drives prices to marginal cost do not seem to consider how strategic pricing in markets with rivalry can lead to quite different types of pricing behavior.

number of participants, the greater the customer's willingness to pay. The leader of AT&T's New York operations, U. N. Bethel, suggested that pricing should be based on value, rather than considerations of cost. Even if it were less expensive to serve business customers, they should be charged a premium price. Bethel explicitly rejected the proposition of basing rates on the cost of service. He pointed out that a new rate schedule had recently been adopted in New York City and the price discrimination embedded in the tariff had stimulated network membership:

> In fixing a [new schedule] we were not governed by the fact that, to some classes of our subscribers, we could furnish service at a less cost than to others, but we were guided largely by the fact that the class of subscribers, the large users, could pay, and would pay without serious objection, the rates which we proposed to charge them. ... [B]y enlarging the potentiality of each of such subscribers, we increase the value of the service to him; and we have taken that into account. ... [T]hat whatever success we have had in recent years, in developing our territory, has been due to the fact that we have charged what the traffic will bear. (AT&T Corporate Archive, January 1900, pp. 211–212)

One year later, Bethel advocated focusing on system benefits, rather than cost-based pricing in a memorandum to the President of AT&T, Frederick Fish. Bethel said that in determining the profitability of a small exchange, the analysis must take into account the associated externalities:

> If the rates in a system embracing several cities and towns of various sizes were based solely on the cost-of-service in each individual locale, the rates in some places would be prohibitive and through having no telephone in such places other places would suffer, possibly to such an extent that the development of them would be seriously interfered with. It is common practice to serve some places at cost and even at below cost, for the general good. There are small towns all through the country where the revenues do not meet the expenses, and the rates have been fixed with the knowledge that they would not yield expenses. This has been done, not because of competition, but for the reason that development in such places has been considered essential to enhance the value of the service in larger places, and to cover the territory. (Bethel to Fish, personal communication, *AT&T Corporate Archive*, 1901)

During the period of its patent monopoly, 1876 to 1894, AT&T had failed to recognize that by encouraging membership of marginal customers, and small towns and villages, the economic potential of the network was not being fully exploited. The entrants' approach to building a network was significantly different. The founder of one of the leading Independent journals noted that "the Bell people worked from the top down and the Independents from the bottom up" (MacMeal, 1934, p. 24). The Independents developed their networks by focusing on the residential market and by introducing service to markets that had been ignored by AT&T. For social and commercial purposes, subscribers in these small towns had a need for telephone access to larger towns. Wholesalers, millers, doctors, and other businessmen who worked in these larger cities realized that their trades would be aided by the establishment of an Independent exchange that could reach markets left undeveloped by AT&T (Allen/Fish, 1903; Johnson, 1939).

In general, the consumer's valuation for access to the network was closely correlated and defined by a hierarchy of demands (Gabel & Weiman, 1997). Distinct market segments were organized according to the utility derived from greater access. These distinct segments fell into two broad categories: business and residential customers (Moyer, 1979). Among the former, the demand for service paralleled the organization's internal structure and external market relations. Core business customers were drawn from larger enterprises engaged in long-distance trades, such as hotels, wholesale merchants, department stores, financial intermediaries, transport companies, and national manufacturers. Typically located in the central business district, these firms demanded telephone connections to keep in touch with distant facilities (such as a branch plant or warehouse), as well as customers and suppliers within their trade area and beyond. Given the value of the information transmitted, core business clients placed an obvious premium on the clarity and reliability of these connections.

Smaller businesses in more residential areas—grocers, drug stores, and tailors—used the telephone less frequently. Moreover, like their customers, they called within a narrower geographic range. The telephone, in conjunction with improved delivery and transport services, enabled many retailers to widen their market scope. They also demanded occasional distant connections to wholesalers in the central business district or a nearby city in order to place orders and arrange deliveries.

The demands of households and smaller businesses differed only by a matter of degree. Economic elites, such as managers and professionals, often used their home telephones for business-related transactions and so valued more extensive connections (Fischer, 1992; Pool, 1979). By contrast, many lower income and working-class households could not afford an individual line, and they either purchased party-line service or frequented public telephones.

At the presidential conference, the relationship between the hierarchy of demand and customers' willingness to pay was clearly stated. Sabin, the President of Bell's West Coast operations, mentioned that when residential customers joined the network, businesses' willingness to pay for service increased: "Every new party line subscriber makes it more necessary for some butcher or grocer or other business place to have a higher priced telephone service" (AT&T Corporate Archive, Telephone Rates-Basis-1880–1908, Box 12, p. 196). Sabin added that residential customers were the "foundation" of the business and that their value had to be considered when designing rates: "I think that we all agree that the foundation of the business is the residence. The cream of the business, of course, is the large manufacturer and the large dealer. But if you get in and get the basis, the residence subscribers, you not only get the cream, but you get all the milk that there is in the coconut" (p. 220). Moreover, by increasing residential membership, profits would be earned on the long-distance network that would not be otherwise. In short, Sabin claimed that the profitability of residential service should not be judged on a stand-alone basis, rather the network benefits should be measured by considering the service's impact on the demand for business and toll service.

The President of AT&T did not attend the 1900 Presidential Conference. Nevertheless, the Letter Books of Frederick Fish, the company's president from 1901 to 1907, demonstrate a shared vision. Fish shared Hall's view that the profitability of a transaction should not be measured by considering the direct costs and revenues obtained by serving a particular subscriber. Fish wrote that regardless of whether or not there was competition, some customers might be served "at a rate so low as not to show a profit. These, however, must be compensated for, either by other subscribers in the same place or elsewhere who pay a substantial profit, or by a profitable toll business based upon the fact that we have a large number of subscribers who use the toll lines" (Fish & Yost, 1903).

Fish would clearly have found odious the alleged market-based pricing rules put forth by Mayo and Kasserman. These two academics have argued that in competitive markets, prices would be driven to cost and would not reflect the value of service. The Massachusetts Department of Public Utilities has taken Mayo's and Kasserman's position to its logical conclusion; i.e., cost-based rates lead to customer class profitability ratios that are equal across a class of customers. Thus, to allow for an earnings differential would entail value-of-service pricing, something that these academics claim is anathema to competitive markets and efficient levels of production. Therefore, the Commission has concluded that the return on investment for residential and business customers should be equalized (Kasserman & Mayo, 1994); Massachusetts Department of Public Utilities, 1990).

When Fish was faced with a proposal to eliminate price discrimination, he summarily dismissed the notion: "Nothing could be more dishonest than charging all business men the same price and charging all residence users the same price, irrespective of the value of the service or the amount of service each one requires" (Fish & Wallace, 1904). Fish felt that the reasonableness or fairness of a rate should not be judged by looking at its price/cost ratio. Rather, he believed that the reasonableness of rates should be judged by considering the overall profitability of the firm. Within this constraint, it made little sense to consider the profitability of a particular service: "The fact is that no one of our rates can be shown to be reasonable or unreasonable apart from the others. The only thing that can be demonstrated, and that in an unsatisfactory fashion, is that, on the whole, one of our companies is making too large, or too little, or just the right amount of money" (Fish & Wheeler, 1906). It is highly instructive to contrast Fish's statement concerning the actual historical pricing behavior of AT&T during an era of competition and no regulation, with the following proclamation:

> In the United States and other western nations, ubiquitous telephone subscribership has been achieved by artificial restraints on the price of basic residential local exchange service. These low prices have been funded by cross-subsidies from other telephone services, such as toll and interconnection charges imposed on other carriers. With the growth in competition for all services, including local exchange, internal cross-subsidies are not only economically inefficient, they distort competitive outcomes and are ultimately not sustainable. (Tardiff, 1996, p. 2)

The so-called artificially low price for residential service was selected by AT&T because it maximized network profits, not because of artificial restraints.

USING COST DATA TO SET TELEPHONE RATES IN AN UNREGULATED TELEPHONE MARKET

The previous section summarized how the President of AT&T and its operating companies believed that rates should be set. The implementation of these concepts was left to the firm's managers. The activity of the middle managers was closely monitored by Fish. Fortunately, some of the details of how rates were designed has been retained. The company's comptroller, W. S. Ford developed rate relationships that reflected the concepts just described.

Ford maintained that even if it cost more to provide residential service, business rates should be higher in order to reflect the commercial value of the service:

> The money actually expended in furnishing service to residences and other small users, cannot fairly be taken directly as a basis for fixing their rates, as there would then be no recognition, in the rate scheme, of the added value of the telephone to the large users, due to the presence, in the Exchange, of the residence and small users. This added value should properly be a basis for assessing the large users a part of the cost of taking on the residences and small users. Recognition of this principle requires low rates to small users, and produces a high ratio of telephones to population. (Ford, 1901)

Ford developed a rate schedule through a three-step process. First, the line, station, and traffic-sensitive expenses were identified. At a minimum, usage rates were designed to recover traffic-sensitive rates. He then took the costs of the loop, the line, and the station, and split them between the costs that were directly attributable to customers and were reflected in the fixed monthly charge, and the costs that he associated with external benefits. Because other customers benefited from increased network membership, it would have been economically irrational to assign all of the loop cost to the end user. In recognition of this externality, the loop cost could be divided evenly between the new customer and other subscribers who might call the new subscriber. Ford added, however, that because there was an option value of connection to the network that was independent of usage, 60% of the loop cost should be recovered from the end user and the remaining 40%, which was an external cost, should be recovered on the basis of those who received the calls (Ford, 1901).

Ford used the volume of incoming calls as the metric for the network externality. According to Ford, for every 10 calls received by one-party business subscribers, two-party business, one-part residential and two-party residential subscribers received seven, two, and one call(s), respectively. These call-receiving values were used to assign the 40% of loop costs that were to be recovered based on value (Ford, 1901).

The data indicate that under Ford's pricing scheme, business customers would pay a premium. His proposal was not well received in all quarters. Bethel wrote to Fish that Ford's proposal gave too much consideration to the cost of service. Bethel advocated that additional emphasis should be placed on pricing that was based on "what the traffic will bear," the method followed by the nation's large railroad companies (Bethel to Fish, personal communication, 1901). Citing Alexander's "Railway Practice," Bethel noted that value-of-service pricing was rampant throughout the economy:

> One of the principal points at issue between theoretical railway reformers and railway managers is, whether freight charges shall be based upon the *cost* of the service rendered, or upon its *value*.
>
> Railroads, in common with authors, doctors, inventors, laborers, lawyers, manufacturers, and most other people who have anything to sell, base their prices upon the value of what they have to offer, rather than upon its cost. Indeed no other basis of railroad services pricing is practicable, as it is by no means the simple matter of calculation it is often assumed to be. The cost of any particular act of transportation cannot even be averaged out, except under the most arbitrary of assumptions.[12]

Unfortunately, AT&T's archives do not indicate the extent to which Ford's proposal was adopted. Regardless, it is clear that value-of-service pricing was a shared belief within this unregulated, competitive telephone company.

---

[12]Ibid. Bethel did not provide the page number for the citation or Alexander's first name.

PRICING PRACTICES OF UNREGULATED FIRMS

The previous discussion should make it abundantly clear that Mayo and Kasserman's perception of how firms operate has little to do with actual business decisions made by firms. Through the use of simplistic economic models, they derive conclusions that have little or nothing to do with the actual behavior of markets in which there is rivalry and economies of scale and scope (on both the demand and supply side). Although this chapter does not suggest that all value-of-service pricing is optimal, it clearly indicates that such behavior is a common occurrence in competitive markets and is looked on by firms as a viable strategic tool to employ in pursuing profitability.

For telecommunications analysts and regulators, the path ahead is obvious. They must move beyond the simplistic models that argue that competition drives price to marginal cost and look instead at the actual pricing behavior of various industries. Businessmen in competitive markets have long realized something that too few utility economists recognize; you want to get customers in the door, not charge a large access fee that acts as a barrier. This is why we have seen, for example, in the credit card industry, the disappearance of access fees.

The mantra that competitive markets drive prices to cost is reasonable to some extent. Where economic profits are being earned, there is an incentive for competitive entry. On the other hand, competitive entry does not mean that each price is set equal to the cost of production. This chapter shows that in markets in which rivalry exists or is threatened, for various strategic reasons, firms do not set price equal to cost. Rather, consistent with recent empirical and theoretical work in economics, if anything, rivalry compels firms to create more elaborate forms of price discrimination.

ACKNOWLEDGMENTS

This chapter is based on my paper, "An Assessment of Universal Service," a report presented on behalf of the Office of the Public Counsel for the State of Florida. I have benefited greatly from the assistance provided by Janet Schloss and Scott K. Kennedy.

REFERENCES

Aley, J. (June 10, 1996). Give it away and get rich: Plus other secrets of the software economy. *Fortune Magazine* [online], p. 4. Available: http://pathfinder.com/fortune/magazine/1996/toc/960610/sop.html

Allen/Fish. (1903, February 16). *Presidential letter books*. New York: AT&T Corporate Archive. AT&T Corporate Archive, *Telephone Rates-Basis-1880–1908*, Box 12.

Anderson, R. H., Bikson, T. K., Law, S. A., & Mitchell, B. M. (1995). Universal access to e-mail: Feasibility and societal implications. Santa Monica, CA: Rand.

Annual Report of the AT&T Company. (1909). (pp. 25–28).

AT&T Corporate Archive. (1900, January). *Telephone service and charges.* (Box 185-02-03).

And a big loss at AT&T, (September 16, 1996). *Business Week,* p. 54.

AT&T Corporate Archive. (1909, January 21). *Dubois to Vail.* (Box 1358).

Ausubel, L. M. (1991, March). The failure of competition in the credit card market. *American Economic Review, 81,* 1, 50–81.

Beilock, R. (1985). Is regulation necessary for value-of-service pricing? *Rand Journal of Economics 16,* 93–102.

Bethel, N. U. to Fish, F.(personal communication, December 24, 1901), in *Telephone Rates-Basis-1880–1908,* Box 12. New York: AT&T Corporate Archive.

Borenstein, S., & Rose, N. (1994). Competition and price dispersion in the U.S. airline industry. *Journal of Political Economy, 102,* 653–683.

Conference Report for the Telecommunications Act of 1996, 104th Congress, 2nd Session, Report 104-458.

Federal Communications Commission. (1939). *Investigation of the telephone industry in the United States* (pp. 129, 135). Washington, DC.

Federal Communications Commission. (1995, November). Reference Book: *Rates, price indexes, and household expenditures for telephone service.*

Fish & Wallace. (1904, November 26). *Presidential letter book* (p. 43). New York: AT&T Corporate Archives.

Fish & Wheeler. (1906, June). *Presidential letter book* (p. 44). New York: AT&T Corporate Archives.

Fish & Yost. (1903, February 5). *Presidential letter book* (p.26). New York: AT&T Corporate Archives.

Fischer, C. (1992). *America's calling: A societal history of the telephone to 1940.* Berkeley: University of California Press.

Ford, W. S. (1901, September 10). Memorandum: Concerning certain peculiar features of telephone exchange service in recognition of which is suggested the system of rates dated March 21, 1901. In *Telephone Rates-Basis-1880-1908,* Box 12. New York: AT&T Corporate Archive.

Gabel, D., & Weiman, D. (1997). Historical perspectives on interconnection between competing local exchange. In D. Gabel and D. Weiman (Eds.), *Opening networks to competition: The regulation & pricing of access* (pp. 75–107). Kluwer Academic Press, Howell, Massachusetts.

Gordon, K., & Taylor, W. R. (1996, April 12). *Comments on universal service.* Attachment to *Comments: In the matter of federal–state joint board on universal service,* Bell South Corporation cc Docket No. 96-45.

In the Matter of The Bell Atlantic Telephone Companies, (March 6, 1995) Tariff FCC No. 10 Video Dialtone Service, Transmittal No. 741, Exhibit A, pp. 4–5. (Affidavit of William Taylor.)

Investigation into Nontraffic Sensitive Cost Recovery, Order No. 18598, 89 PUR4th 258, Florida Public Service Commission, 265–66 (December 24, 1987).

Johnson, F. G. (1939). Experience of a pioneer physician in northern Wisconsin. *Wisconsin Medical Journal, 38,* 580.

Kahn, A. (1984). The road to more intelligent telephone pricing. *Yale Journal on Regulation 1,* 139, 142–145, 155.

Kahn, A., & Shew, W. (1987). Current issues in telecommunications regulation: Pricing. *Yale Journal on Regulation, 4,* 191, 202.

Kasserman, D. L., & Mayo, J. W. (1994). Cross-subsidies in telecommunications: Roadblocks on the road to more intelligent telephone pricing. *Yale Journal on Regulation, 11,* 119, 127–130, 135.

MacMeal, H. (1934). *The story of independent telephony.* Chicago: (Independent Pioneer Telephone Association).

Mandell, L. (1990). *The credit card industry: A history.* Boston: Twayne Publishers.

Massachusetts Department of Public Utilities. (1990, June 29). *Propriety of the Rates and Charges of New England Telephone.* (Docket 89-300, pp. 10–12, 16, 21).

Morris, B. (1996, March 4). The brand's the thing. *Fortune Magazine* [Online], p. 5. Available: http://pathfinder.com/fortune/magazine/1996/toc/960304.htmc

Motorola goes for the hard cell. (1996, September 23). *Business Week,* p. 39.

Moyer, A. J. (1979). *Urban growth and the development of the telephone: Some relationships at the turn of the century.* In I. Pool (Ed.), *Social impact of the telephone.* (pp. 357–365). Cambridge, MA: MIT Press.

Nagle, T. T. (1987). *The strategy and tactics of pricing: A guide to profitable decision making.* Englewood Cliffs, NJ: Prentice-Hall.

Organization for Economic Co-Operation and Development (OECD). (1995, May 15). *Mobile communications: Pricing strategies and competition.* 80–97.

Pool, I. de Sola. (1979). *Foresight and hindsight: The case of the telephone.* In I. Pool, (Ed.), *The social impact of the telephone* (p. 142). Cambridge, MA: MIT Press.

Response of Southern New England Telephone. (1995, October 13). TE052 Supplemental response. Application of the Southern New England Telephone Company for financial review and proposed framework for alternative regulation.

Scotchmer, S. (1985). Two-tier pricing of shared facilities in a free-entry equilibrium. *Rand Journal of Economics, 16,* 456–472.

Sprout, A. L. (1995). The rise of Netscape. *Fortune Magazine* [Online], p. 2. Available: http://pathfinder.com/fortune/magazine/1995/950710/impotech/netscape/netscape.html

Stigler, G. (1987). *Theory of Price.* New York: Macmillan.

Talley, W. (1989). Joint cost and competitive value-of-service pricing. *International Journal of Transport Economics, 16,* 119–130.

Tardiff, T. J. (1996, July). *Universal service with full competition.* Paper presented at The Telecommunications Universal Service Symposium, Wellington, New Zealand.

The card: A work in progress. (1996, September 18). *The New York Times,* p. D1.

The software revolution. (1995, December 4). *Business Week.*

Thomsen, K. A. (1987). *The Global strategy of the Gillette Corporation.* Unpublished masters thesis, Massachusetts Institute of Technology, Boston.

U.S. Department of Commerce, Bureau of the Census. (1907). *Telephones,* 74–75, 80.

Vietor, R. (1994). *Contrived competition: Regulation and deregulation in America.* Cambridge, MA: Harvard University Press.

Wisconsin Telephone News 1,1 (December 1906)

10

Universal Service: A Stakeholder Response

James C. Smith
Ameritech

Although the speakers at the New York Law School Conference have certainly had differing opinions on how universal service goals should be viewed going forward, there seems to be a great deal of understanding and agreement on how universal service has evolved to the current situation. At Ameritech, we see the very real prospect of imminent competition in nearly all of our markets. Unfortunately, as things stand currently, we also see the very distinct and dangerous possibility that as this competition develops, we continue to bear the burdens of sustaining current universal service policies in the territories we serve. However, current policies are not sustainable—the burden is one that Ameritech cannot continue to bear alone as it faces competition throughout the business. With time running out on current universal service policies, Ameritech is actively developing solutions that are sustainable, workable, fair to customers, and, perhaps most important, politically doable in this future environment. Therefore, reflecting on some of the issues discussed earlier, I describe Ameritech's view on what needs to be done to preserve the essential components of universal service in the fully competitive environment that is on us.

There appears to be a general understanding that current universal service policies in this country are the result of both rational economic decisions and the very heavy influence of political processes. As a telecommunications service provider, there has been an economic incentive for Ameritech to maintain and increase subscribership because doing so increases the value provided to all customers. The means for enhancing subscribership include providing assistance to low-income customers, to customers residing in costly to serve areas, and to customers with special needs, such as the hearing-impaired. This assistance consists of reductions in monthly rates or nonrecurring charges, more lenient deposit requirements or past due bill payment arrangements, free access to services that help control costs such as toll-blocking services, and just plain better communications and accessibility.

Providing special assistance to these types of customers not only has an underlying economic rationale, but also aligns well with the expectations of the political process so long as any resultant subsidies are implicit, that is, hidden from widespread public scrutiny. Unfortunately, however, the influence of political processes has expanded the goals inherent in our current universal service policy well beyond the realm of economically justifiable objectives. The goals of keeping all residential rates priced as low as possible, regardless of the impact on rates charged to other customers, and of keeping rates averaged regardless of the underlying cost differences have resulted in major pricing anomalies—that is, current prices are far different from what would have evolved under market conditions free from these political influences. As a result, businesses have been required to pay far more than a market-based price so that residential services could be priced low. Urban subscribers pay prices well above cost so that service to their sometimes more affluent suburban and rural counterparts can be subsidized. Furthermore, users of so-called discretionary services, such as toll and special features, have paid large premiums in order to fund below market prices for so-called essential basic services.

Although the manipulation of prices has been the most obvious tool used by regulators to achieve the full breadth of politically influenced universal service goals, other regulatory actions have also had significant impacts. For example, the rather predictable regulatory practice of consistently underestimating the true economic depreciation that was occurring on in-service assets provided an obscure way to keep overall prices lower in the short run. Subsidies were generated for current customers to be paid for by future customers. In another vein, regulators imposed readiness to serve requirements and associated penalties for noncompliance that encouraged providers to err on the plus side when estimating how many facilities were required to serve newly developing areas. Such requirements ensured that no customer had to wait long to be connected to the network; but they also imposed significant, hidden costs on all customers.

There also appears to be a general recognition that these traditional mechanisms for implementing universal service policies could succeed only in the closed system of local telephone service monopolies that characterized our industry for many decades. As an aside, this premise is strongly supported by anecdotal evidence gathered from individuals recently exposed to the vagaries of pricing in our industry. Ameritech, for example, has recently hired many new managers from unrelated competitive industries to build the marketing and customer service skills that are required as competition enters our markets. These new managers are absolutely amazed once they come to understand our current system of price subsidies and cost shifting, for it resembles nothing they have experienced in the competitive industries from whence they came.

The closed system of monopoly allowed regulators to, on the one hand, manipulate price structures, distort the true economic costs of capital, and impose noneconomic legal requirements on providers, and, on the other hand, keep companies on a sound financial footing in order to fulfill legal mandates and to attract the investor capital necessary to keep pace with growth and the rapid pace of technical change. The financial safety net extended to providers in this monopoly environment made them willing, albeit often reluctant, partners with regulators in carrying out the political agenda.

However, the closed system of monopoly local providers is no longer a reality. At the fringes, it has been gone for a long time—ever since the ability to at least partially bypass the networks of the local providers with private microwave systems or fiber-optic networks built by competitive access providers became a reality for large customers more than 10 years ago. Now the local monopoly is being attacked at its core as multiple providers are being authorized to provide a full panoply of competitive local services.

The mechanisms designed to promote and perpetuate universal service goals in the old days of a monopoly system are simply not sustainable in the new, full-blown competitive environment. Business customers will not voluntarily continue to pay an implicit tax to benefit residential customers if they can avoid that tax by giving their business to an alternative provider. Similarly, urban customers, who will surely be the first beneficiaries of competitive entry, cannot be expected to voluntarily continue to subsidize the high costs of suburban and rural areas. So, once full competition becomes a reality, the current universal service system begins to rapidly unravel.

Does that mean that we must abandon our commitment to universal service as we embrace competition in our industry? No, but local competition does mean that there needs to be an evolution from the current system of implicit subsidies and unilaterally imposed requirements to a new system based on explicit assistance, competitively neutral funding mechanisms, and symmetrical application of regulation. There should also be a comprehensive review and assessment of the universal service goals that evolved during the decades of a closed monopoly system, leading to revised, streamlined goals appropriate for a competitive industry. A transition plan is also needed for moving from where we are today to where we need to be in the future.

It is incumbent on regulators and government policymakers to reach these important policy decisions and to communicate them to all industry participants early in the process of opening local markets to competition. Failure to do so is likely to create misleading incentives for new participants, resulting in inefficient entry and wasted investments. Introduction of local competition without a viable long term universal service plan that both reassesses the goals and revamps the system to achieve the goals in a competitively neutral manner will also be strongly resisted by even the most progressive incumbent providers, leading to gridlock and unnecessary delay in bringing the full benefits of competition to customers.

A comprehensive review and assessment of traditional universal service goals requires a three-step approach, starting with the articulation of each and every goal that has guided past actions. In this regard, clarifying the difference between a goal and the means of achieving the goal is important. For example, is rate averaging the goal, or is it simply a traditional mechanism used in the past to achieve the goal of reasonable prices?

The second step is to determine which goals will continue to require regulatory intervention. The entry of competition into local telephone markets will bring about a marketplace discipline that may supplant the need for regulators to mandate certain practices. For example, the widespread practice of mandating the availability of flat-rate local calling service will no longer be necessary when competitors begin vying for customer business because, in order to succeed in the marketplace, firms will have to offer service packages that fulfill customer needs.

The third step is the most difficult part—taking each of the remaining historical goals and making an evaluation of whether the future benefits outweigh the costs of regulatory intervention. This is a difficult step to acknowledge for the traditional regulatory mindset, for it requires tacit recognition of the principle that there is no such thing as a free lunch. However, it is a step that must and should be taken.

Done properly, assessing each historical goal from a cost–benefit standpoint is likely to result in a major scaling back of traditional universal service objectives. On the one hand, the goal of maintaining and perhaps even increasing subscribership is likely to survive, for, if assistance is appropriately targeted and based on need, this goal can be achieved at a relatively low cost. On the other hand, the achievement of most other goals probably comes at far too large a cost. Studies abound that estimate significant, societal benefits from liberating this increasingly important segment of our economy from the burden of obtrusive government oversight to the discipline of the competitive marketplace.

This scaling back of traditional universal service goals may not be as politically difficult as it may seem. Experiences in other states can be enlightening. During the 1980s, for example, Illinois went through a major transformation of its universal service goals. The principle of rate averaging was basically rejected, and the historical subsidy flows described earlier were reduced considerably. Customers in higher cost suburban areas were asked to pay higher prices, and customers in even higher cost rural areas were asked to pay even higher prices. Prices in urban areas, on the other hand, were reduced. The prices for making a toll call were aligned with the prices for making local calls; and the disparity between business rates and residential rates was dramatically reduced.

As a result of this price restructuring, there was no measurable change in telephone household subscribership. Furthermore, there was virtually no negative reaction to these dramatic changes. A few groups, purporting to represent consumers, voiced moderate objections; but very few customers complained. Overall, there was virtually no political fallout. In addition, these changes permitted Illinois to remain an overall low price leader and paved the way for Illinois to be one of the first test beds for the efficient introduction of competition.

I am optimistic that policymakers are beginning to see the value of scaling back governmentally imposed universal service goals in these emerging competitive markets. In Michigan, for example, the recently enacted state telecommunications law *mandates* the rebalancing of rates so that, by the year 2000, no service can be priced below its economic cost. The new Michigan law also specifically permits rate deaveraging. In Michigan, policymakers evidently believe that eliminating implicit subsidies is the best way to evolve to a fully competitive market. Furthermore, in both Illinois and Michigan, policymakers wisely decided to adopt 3- to 5-year transition plans for phasing in these changes, giving customers time to understand the impacts and to make necessary adjustments.

What remains after this process of thorough review and assessment of historical universal service goals is the need to develop long-term sustainable processes to achieve the (hopefully) reduced and revised list of new objectives. Transitional issues associated with migration to the new universal service goals must also be addressed. This requires explicit identification of the subsidies to be retained, development of competitively neutral mechanisms for funding these subsidies, and symmetric application of regulation to all providers.

Ameritech believes that the avoidance of asymmetric regulation of providers is one of the keys to the sustainability of any universal service goals. Asymmetric regulation encourages creamskimming—that is, the selective provisioning of service by new entrants to more highly profitable customers—and handicaps the ability of incumbents to respond to market forces. Under such circumstances, the ability of incumbents to continue to serve as a carrier of last resort is put in severe jeopardy.

The achievement of symmetric regulation can mean either subjecting new providers to the same rules and obligations of existing providers or relieving existing providers of the same. Practically, a balance of both approaches is required.

There will also be situations where neither approach is viable in the short run. For example, under current regulation, most incumbent providers serve as a carrier of last resort and thereby provide service to some customers below cost. Because at least in the short run it is neither feasible to relieve the incumbent of these burdens nor tenable to impose similar obligations on new providers, transitional mechanisms are required to maintain competitive balance.

Ameritech has been a national leader in developing and advocating comprehensive plans for the introduction of competition and in seeking to ensure that universal service goals are not completely lost in the rush to make local competition a reality. Through groundbreaking proposals such as the regional Customers First Plan and individual state initiatives, Ameritech's efforts in this area have helped to draw attention to the intricate interrelationships among all of the various initiatives being considered by policymakers.

The resolution of each of the many issues pending in the industry today—including complex proposals to unbundle network components and to offer wholesale discounts to resellers—will have important implications for the sustainability of universal service objectives going forward. It is critical for policymakers to integrate the resolution of all of these issues, and to resist the temptation to resolve them on a piecemeal basis.

In summary, three key points have been made. First, we should not delude ourselves into thinking that the current processes for achieving universal service goals can be sustained in the new competitive environment. They cannot.

Second, we should be open-minded in fundamentally rethinking the current universal service policies and the means to accomplish them in a competitive environment. Major reform is the easiest and, in the long run, may be the least painful means of ensuring the rapid introduction of competitive benefits throughout all of our markets.

Third, we should ensure that the processes established for achieving true universal service objectives in the future are sustainable by designing rules that are symmetric among providers in the long run and, if asymmetric in the short run, by providing appropriate transitional measures.

V

EMBARKING ON A NEW UNIVERSAL SERVICE POLICY: THE ROLE OF THE FEDERAL GOVERNMENT

11

Review of Federal Universal Service Policy In the United States

Barbara A. Cherry
Steven S. Wildman
Northwestern University

Through passage of the Telecommunications Act of 1996, Congress altered radically the federal framework for regulating the telecommunications industry. As part of this framework, Congress also codified universal service policy. Although there is some debate as to whether Congress contemplated a policy of universal service under the original Communications Act of 1934 (Mueller, 1993), there is no doubt that universal service is a critical component of the new regulatory regime set forth in the Act.

Because other chapters of this book assume that the reader has familiarity with universal service policy established in the Act, this chapter provides an overview of the key provisions of universal service in the Act and their implementation to date by the Federal Communications Commission (FCC). This overview, however, is intended to be descriptive only and to serve as background for the arguments and prescriptions offered in the other chapters. It is also meant to complement the discussion of universal service policy from the perspective of the states provided in chapter 14 by Thomas Bonnett.

GENERAL FRAMEWORK OF UNIVERSAL SERVICE POLICY

In the Act, universal service policy is stated with great specificity, although many aspects are necessarily left for the FCC and State Commission implementation. Given that the essential elements of Congress' universal service policy are contained primarily in sections 254 and 214(e) of the Act, this chapter is based on these sections. However, it should be kept in mind that a fuller appreciation of the role of universal service in the overall regulatory framework does require familiarity with the other provisions of the Act as well.

Increased Reliance on Explicit Funds

In recent decades in the United States, universal service has meant the availability of basic telephony service at reasonable rates to all citizens (Mueller, 1993). This goal has been pursued largely through the use of implicit support flows among classes of services and/or customers. Although there is debate as to whether these support flows are subsidies in the true economic sense (Aron & Wildman, 1996; Gable, chap. 9, this volume; Panzar & Wildman, 1995), there is agreement that regulators have manipulated the price structure to achieve social goals, such as those embodied in the concept of universal service.

To a lesser extent, government has relied on the establishment of explicit funds or programs, such as the Lifeline and Link-Up programs, to assist targeted groups of customers. The hallmark of the policy framework set forth in section 254 is the increased reliance on explicit funds to support universal service for targeted groups of end-users. However, such increased reliance on explicit support is tempered by various statutory restrictions on rate deaveraging or rate rebalancing. For example, service rates in rural and high cost areas must be reasonably comparable to those in urban areas (section 254(b)(3)), and rates for interstate interexchange services must be the same across States (section 254(g)). The framework of section 254 is represented pictorially in Fig. 11.1.

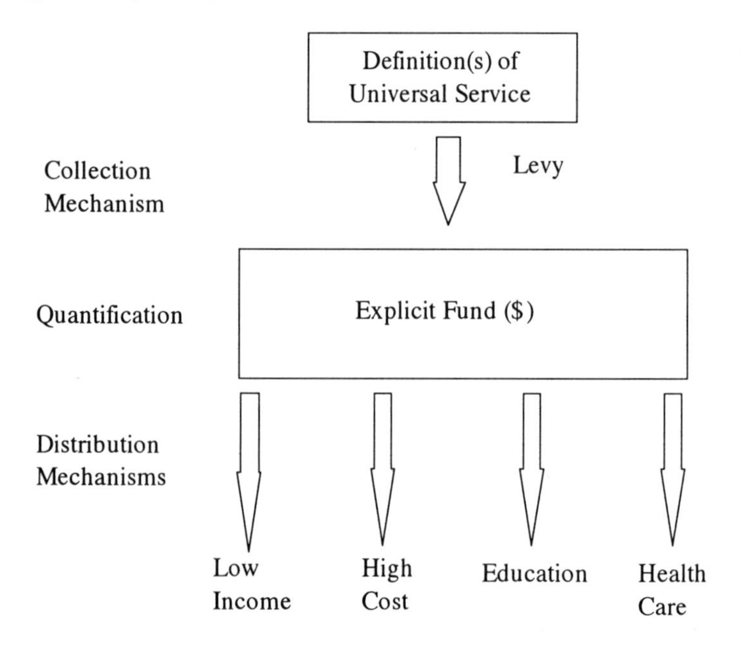

FIG. 11.1 Framework of Section 254 of TA96.

The logic of section 254 is the creation of definition(s) for universal service, the identification of specific groups of end-users for which the availability of universal service is to be assured, the establishment of explicit universal service fund(s), quantification of amounts for the explicit fund(s), collection mechanism(s) for raising money for the explicit fund(s), and distribution mechanisms for distributing funds for the benefit of targeted groups of end-users. The targeted groups of end-users are identified in section 254 as low-income customers, customers living in rural (high cost) areas, certain educational institutions and libraries, and health care providers serving customers in rural areas.

The definition of universal service varies with the end-user group being targeted. For example, as discussed later, the definition of universal service for educational institutions includes all commercially available telecommunications services, whereas, for residential customers, it is restricted to certain voice grade services. Similarly, the quantification and distribution of support will also vary by end-user group. However, (for a given jurisdictional level) the same collection mechanism is to be used—a levy assessed on all telecommunications service carriers, as defined under the Act—for raising the funds for the explicit fund.

Dual Jurisdiction Retained

Congress did retain the dual jurisdictional nature of telecommunications regulation. The FCC still has jurisdiction over the provision of interstate services and the states retain jurisdiction over the provision of intrastate services. Therefore, the preceding framework applies to both federal and state policies, and the necessary coordination between them.

An important mechanism for coordinating federal and state universal service policies is the FCC's implementation of the recommendations of the Federal–State Joint Board on Universal Service created under section 254. The Federal–State Joint Board issued its recommendations regarding the financing and management of a federal universal service support fund in November 1996 (*Recommended Decision*, 1996). The FCC established rules (*Universal Service Rules*, 1997) responsive to these recommendations in its *Report and Order* (hereinafter referred to as the *FCC Report and Order*) issued on May 8, 1997, a summary description of which is provided later.

The states also have the authority to impose requirements to preserve and advance universal service using the same framework. Thus, action by a state may also be undertaken for purposes of pursuing state universal service goals, including the establishment of a state universal service support fund. However, there are limits on state action. Section 253(d) of the Act provides the FCC with preemption powers under certain circumstances. Furthermore, under section 254(f), state regulations must not be inconsistent with FCC rules. In this regard, according to the *FCC Report and Order* (1997), states may adopt additional regulations and standards to preserve and advance universal service, but, "only to the extent that such regulations adopt additional specific, predictable, and sufficient mechanisms to support such definitions or standards that do not rely on or burden Federal universal service support mechanisms" (section 254(f)).

Competitive Neutrality Principle

Section 254 (b) states several principles on which the Joint Board and the FCC are to base their policies. These principles include, but are not limited to: the availability of quality services at just, reasonable, and affordable rates; that federal and state mechanisms be specific, predictable, and sufficient; and that all providers of telecommunications services make equitable and nondiscriminatory contributions to the preservation and advancement of universal service.

Under section 254(b)(7), additional principles may also be established by the Joint Board and FCC. An additional principle of competitive neutrality was recommended by the Joint Board and adopted by the FCC. "Universal service support mechanisms and rules should be competitively neutral. In this context, competitive neutrality means that universal service support mechanisms and rules neither unfairly advantage nor disadvantage one provider over another, and neither unfairly favor nor disfavor one technology over another" (par. 47). Thus, *both* federal and state universal service support mechanisms must be competitively neutral.

Competitive neutrality of state universal service policies is also mandated by Congress in section 253(b). Violation of competitive neutrality is a basis for FCC preemption of a state or local government statute, regulation, or legal requirement under section 253(d). Thus, competitive neutrality is required by both the Act and the *FCC Report and Order.*

Carriers Eligible to Receive Support

A final element of the general universal service framework is the distribution of support for the benefit of various targeted end-user groups. Congress provided that universal service support be made available to "eligible carriers," as defined under section 214(e)(1). Section 214(e)(1) provides in relevant part:

> A common carrier designated as a eligible telecommunications carrier ... shall, throughout the service area for which the designation is received—
>
> (A) offer the services that are supported by Federal universal service support mechanisms under section 254(c), either using its own facilities or a combination of its own facilities and resale of another carrier's services (including the services offered by another eligible telecommunications carrier); and
>
> (B) advertise the availability of such services and the charges therefore using media of general distribution.

Generally, eligible carriers are designated by state commissions (section 214(e)(2)). However, if no carrier is willing to provide universal service supported by federal universal service support mechanisms to an unserved community or portion thereof, the eligible carrier is to be designated for interstate purposes by the FCC and for intrastate purposes by the state commission (section 214(e)(3)).

Under section 254(e), only common carriers satisfying this definition of an eligible carrier are eligible to receive federal universal service support. However, there is an exception. Under section 254(h)(1)(B), all telecommunications carriers are required to provide certain educational institutions and libraries with telecommunications services at a discount from that charged for similar services to other parties. Notwithstanding section 254(e), *any* telecommunications carrier providing such discounts shall either have the discounted amount treated as an offset to its obligation to contribute to universal service support or receive reimbursement for that amount through universal service support mechanisms.[1]

IMPLEMENTATION OF UNIVERSAL SERVICE BY THE FCC

In order to implement the general framework provided by Congress in the Act, the FCC issued its *Report and Order* in Docket no. 96-45 on May 8, 1997. Given limited space here, it is not possible to describe all the details of the FCC's Order, much less its nuances.[2] However, highlights of the Order are described. Elsewhere in this volume, authors of the various chapters provide more detail as to those provisions relevant to their chapters' discussions.

Definition of Universal Service

Rural, Insular, and High-Cost Areas. Pursuant to section 254(c)(1), the FCC established a definition of supported services for rural, insular, and high-cost areas. This definition is set forth in Rule 54.101 of Part 54, Title 47 of the Code of Federal Regulations (CFR). In summary, the following services and functionalities are included: voice grade access to the public-switched network; some (unspecified) amount of local usage; dual tone

---

[1]According to the FCC's definition of universal service for eligible educational institutions and libraries, some *nontelecommunications carriers* may also be eligible to receive support for providing Internet access or internal connections.

[2]The FCC's *Report and Order* exceeds 800 pages, including appendices. The rules alone exceed 40 pages. Throughout this chapter, paragraph numbers designate paragraphs of the *Report and Order* and rule numbers refer to the Universal Service Rules set forth in Appendix I of the *Report and Order*.

multifrequency signaling or its functional equivalent (e.g., touch-tone service); single-party service or its equivalent; and access to emergency services, operator services, interexchange service, and directory assistance. This definition is used for purposes of determining universal service support for residential and single-line business customers in high-cost areas.

Low-Income Customers. For purposes of providing support to residential, low-income customers, the definition of universal service for rural, insular, and high-cost areas is used but is augmented to include toll limitation. Toll limitation is defined in rule 54.400 as consisting of both toll blocking and toll control. Toll blocking does not allow the completion of outgoing toll calls, whereas toll control allows consumers to specify a certain amount of toll usage that may be incurred on their telecommunications channel per month or per billing cycle.

Educational Institutions and Libraries. Educational institutions and libraries benefit from a much broader definition of services supported by universal service mechanisms. Rule 54.502 provides that supported services include all commercially available telecommunications services. Furthermore, rule 54.503 states that supported services include Internet access and the installation and maintenance of internal connections. In its *Report and Order,* the FCC elaborated by stating that internal connections include inside wire (pars. 451–456) , as well as routers, hubs, network file servers, and wireless local networks, but not personal computers (pars. 459–463).

Health Care Providers. For health care providers serving persons residing in rural areas, Rule 54.613 provides that supported services include commercially available telecommunications service using a bandwidth capacity of 1.544 Mbps or less. In addition, such telecommunications services as are available in the relevant urban area (to be determined by the FCC) are to be supported and made available to eligible health care providers in rural areas.

The Collection Mechanism

In section 254(b)(4), Congress states that the mechanism for collecting universal service funds should consist of equitable and nondiscriminatory contributions from all telecommunications service carriers, as defined under the Act. For purposes of a federal fund, authority is granted to the FCC to require contributions from carriers providing interstate telecommunications services (sec. 254(d)). In applying section 254(d), the FCC decided that the definition of a carrier providing interstate telecommunications services be construed broadly by including both wireless and wireline providers. However, in its *Report and Order,* the FCC clarified that some entities are not carriers and are not required to make contributions, such as Internet and enhanced service providers (par. 788) as well as private network operators that do not lease excess capacity (par. 786).

The FCC also defined the revenue base against which contributions from telecommunications carriers for federal universal service support mechanisms are to be levied. First, as to jurisdiction, rule 54.703(b) states that, to support educational institutions and health care providers, contributions are to be levied on both the interstate and intrastate revenues of interstate carriers. However, under rule 54.703(c), contributions to support low-income and high-cost customers are to be levied only on the interstate revenues of interstate carriers, including revenues from the international services billed to domestic end users.

Second, as to the type of revenue base, rule 54.703 provides that contributions be levied on end-user revenues. In its *Report and Order*, the FCC clarified that enduser revenues include revenues derived from telecommunications services (par. 844) and from the subscriber line charge (par. 844), but exclude payments received from the universal service support fund (par. 857).

In basing contributions to federal universal service support on end-user revenues, the FCC expressly declined to adopt the Joint Board recommendation that contributions be based on telecommunications revenues net of payments to other carriers because carriers may differ in their ability to recover their contributions from customers (par. 850). However, at the same time, the FCC declined to mandate that carriers recover contributions through an end-user surcharge (par. 853).

Quantification and Distribution of Universal Service Support

In its Order, the FCC quantified numerous aspects of federal universal service support. For some targeted end-user groups, actual dollar amounts are specified. For others, the methodology for quantification is either specified or deferred to a later date.

Low-Income Customers. Low-income customers will continue to receive universal service support under two mechanisms—Lifeline service and Link-Up assistance. Lifeline service is the provision of retail local service, which includes the services and functionalities enumerated in rule 54.101 described earlier. Under rule 54.401, carriers may not disconnect Lifeline service for non-payment of toll charges, may not collect a service deposit in order to initiate Lifeline service if the customer voluntarily elects toll blocking, and must provide toll limitation to Lifeline customers at no charge.

Under rule 54.403, baseline Lifeline support is provided from the federal fund in an amount equal to $3.50 per qualifying low-income customer per month. If a state commission approves an additional reduction in the amount of $1.75 per month paid by Lifeline customers, additional federal Lifeline support in the amount of $1.75 will be made available to the carrier providing Lifeline service to that customer. Further additional federal Lifeline support in an amount of one half the amount of any Lifeline support approved by a state commission will also be made. However, total federal Lifeline support shall not exceed $7.00 per qualifying low-income customer. Under rule 54.407, all the preceding amounts are paid directly to the eligible telecommunications carrier providing Lifeline service to qualifying low-income customers.

Link-Up is an assistance program for qualifying low-income customers that reduces the carrier's customary charge for initiating telecommunications service for a single connection at a consumer's principal place of residence. Under rule 54.411, the qualifying low-income customer pays half of the customary charge or $30.00, whichever is less, and is permitted to pay for the charges that are assessed for commencing service on a deferred schedule and without interest. Interest charges not assessed to the customer shall be for connection charges in amount of up to $200.00, and deferral is to be for a period not to exceed 1 year. Under rule 54.413, eligible telecommunications carriers may receive reimbursement from federal universal service support funds for the revenue foregone in reducing their customary charge for commencing service and for providing a deferred schedule for payment of the charges assessed under the Link-Up program.

Educational Institutions and Libraries. Under rule 54.507, an annual cap on federal universal service support for schools and libraries is set at $2.25 billion per funding year. However, no more than $1 billion shall be spent for the period from January 1, 1998 through June 30, 1998. Up to one half of unused funds may be carried over to a succeeding calendar year.

Although this annual cap on funds for schools and libraries is easy to describe, the methodology by which this amount is allocated among eligible institutions is not. Implementation of this aspect of federal universal service support is provided in Rules 54.500 through 54.517, which are briefly summarized here.

As previously discussed, telecommunications carriers are to provide telecommunications services at discounts. Under rule 54.504, schools and libraries are required to seek competitive bids for all such discounts. In turn, a ceiling for a carrier's bid is the lowest corresponding price for similar services charged to similarly situated, nonresidential customers in the same geographic area.

Rule 54.505 provides a matrix by which discounts, ranging from 20% to 90%, are applied to the competitively bid prices received by eligible schools and libraries. The varying discounts are allotted to schools and libraries based on indicators related to the levels of poverty and the cost of providing telecommunications service in the geographic areas where they are located. Federal universal service support is provided for both the provision of interstate and intrastate services to schools and libraries; however, support for intrastate services is contingent on the state's establishment of intrastate discounts at levels no less than those applicable for interstate services (rule 54.506(e)).

For purposes of seeking bids for service, schools and libraries are permitted to participate in consortia—including other eligible schools and libraries, eligible health care providers, and public sector governmental entities—in order to aggregate their demands (rule 54.501(d)(1)). However, only eligible schools and libraries in such consortia may receive service at the discounted rates (rule 54.501(d)(2)), and eligible services provided at a discount may not be sold, resold, or transferred for consideration (rule 54.513(a)).

Utilizing this methodology, eligible schools and libraries pay the discounted rates to carriers. The difference between the competitively bid price and the discounted price is then treated as an offset to the carrier's obligation to provide contributions to the federal universal service fund—although direct reimbursement from the fund is provided to the carrier if the offset exceeds the carrier's obligation to the fund (rule 54.515). If the total annual fund is exhausted, schools and libraries then pay the competitively bid, but nondiscounted, prices. But, rules of priority are established to ensure that funds are used for the benefit of the most economically disadvantaged schools and libraries (rule 54.507(f)).[3]

Health Care Providers. Under rule 54.623, the FCC established a cap of $400 million per funding year from the federal universal service fund for health care providers serving persons in rural areas. There are two methodologies by which this funding is to be used to support services to health care providers.

First, telecommunications carriers must charge eligible health care providers an urban rate for each supported service (described earlier), which is defined as a rate that is no higher than the highest tariffed or publicly available commercial rate for a similar service in the closest city in the state with a population of 50,000 or more people, taking distance charges into account (rule 54.605). The rural rate is defined as the average of the rates actually being charged to commercial customers for identical or similar services in the rural area in which the health care provider is located (rule 54.607). The identity of the carrier and the most cost-effective, commercially available telecommunications service to be provided to the eligible health care provider are determined by a competitive bidding process (rule 54.603); and the difference between the rural and urban rate of the provided service is treated as an offset against the carrier's federal universal service support obligation (rule 54.611). However, if the offset exceeds the carrier's obligation, then the carrier receives reimbursement from the fund in the amount of the excess.

---

[3]With appropriate modifications, the preceding framework also applies to covered services—including Internet access and installation and maintenance of inside wire—provided by nontelecommunications carriers (rule 54.517).

Second, eligible health care providers who lack toll-free access to an Internet service provider (ISP) may receive the lesser of the toll charges incurred for 30 hours of access to an ISP or $180 per month in toll charge credits to be applied against toll charges for connecting to an ISP (rule 54.621). As with the previously described discounts, the amount determined under rule 54.621 is treated as an offset—or reimbursement, if the offset exceeds the carrier's obligations—to the carrier's federal support fund obligation.

As with eligible educational institutions and libraries, health care providers may join to form consortia with other eligible health care providers, with schools and libraries, and with public sector entities for the purpose of aggregating demand in ordering telecommunications services (rule 54.601(b)). In addition, supported telecommunications services provided to eligible health care providers may not be sold, resold, or transferred for consideration (rule 54.617).

Customers in High-Cost Areas. The category of customers for which the largest amount of universal service support will be provided consists of customers residing in high-cost areas. However, quantification of support for high-cost areas has been left largely unresolved in the FCC Order.

The FCC did find that the cost methodology to be used to calculate the cost of providing universal service for high-cost areas should be based on forward-looking economic cost (par. 223). However, the FCC has not adopted a specific cost model at this time. The FCC found the proffered cost models, such as those presented by participants in the proceeding, to not yet be sufficiently reliable (pars. 232–245). Therefore, selection of a model and related issues—such as the size of the serving area for which costs are to be estimated for purposes of calculating support requirements—are deferred to further rulemaking.

Furthermore, rural and nonrural carriers are to be treated differently. Rural carriers are to continue to receive support based on their embedded cost using current support programs (Weinhaus et al., chap. 8, this volume) with some modifications, for at least three years. Thereafter, a transition will be made to a forward-looking cost methodology (par. 294). For nonrural carriers, a specific forward-looking cost model is to be adopted by August 1998 through a further rule-making proceeding (par. 245). In this regard, so long as certain criteria for a forward-looking cost model are met, a state may select either its own cost model or use the one adopted by the FCC to determine the amount of federal universal service support for carriers in that State (par. 250). Support to nonrural carriers based on the adopted models will not commence until January, 1999 (par. 245).

The FCC also established a methodology for setting revenue benchmarks to which costs will be compared to determine the amount of support that a carrier will receive for serving high-cost areas. Nationwide benchmarks are to be based on average revenues per line for local, discretionary, interstate, and intrastate access services, and other telecommunications revenues (pars. 259–263). Two separate benchmarks are to be adopted, one for residential lines and another for single-line businesses. Although precise calculation of the level of the benchmark is deferred in order to be consistent with the forward-looking cost methodology yet to be adopted, current data suggest that the benchmarks are likely to be approximately $31 per month for residential lines and $51 per month for single-line businesses (par. 267).

Administration

Finally, under rule 54.701, the FCC adopted the Joint Board recommendation that a neutral, third-party administrator be selected to administer the federal universal service support mechanism. In particular, the FCC will create a Federal Advisory Committee, whose sole responsibility will be to administer a competitive bid process for selecting and nominating a neutral third party for FCC consideration. Until such time as that party can be selected, the National Exchange Carrier Association (NECA) is appointed the temporary administrator.

CONCLUSION

The preceding description highlights essential elements of universal service policy as established by Congress in the Telecommunications Act of 1996 as well as the early stages of its implementation by the FCC. Although this description belies the complexity of the issues faced and discussed by the FCC in its May 8, 1997, *Report and Order,* it should serve as a useful guide to the broad contours of federal universal service policy. It should also help contextualize references to the Act and federal universal service policies in other chapters in this volume.

REFERENCES

Aron, D., & Wildman, S. (1996). The pricing of access in telecommunications. *Industrial and Corporate Change, 5,* 1029–1048.

FCC Report and Order. In the Matter of Federal-State Joint Board on Universal Service, CC Docket No. 96-45 (1997).

Gable, D. (1998). Recovering access costs: The debate. Chapter 9, this volume.

Mueller, M. (1993). Universal service in telephone history: A reconstruction. *Telecommunications Policy, 17,* 352–369.

Panzar, J., & Wildman, S. (1995). Network competition and the provision of universal service. *Industrial and Corporate Change, 4,* 711–720.

Recommended Decision. In the Matter of Federal-State Joint Board on Universal Service, 12 FCC Rcd 87 (1996).

Telecommunications Act of 1996, Pub. L. No. 104-104 , 47 U.S.C.A. Sec. 151 *et seq.* (West Supp. 1997).

Universal Service Rules, in Appendix I, FCC Report and Order, In the Matter of Federal-State Joint Board on Universal Service (1997) (to be codified at Part 54, Title 47 of the Code of Federal Regulations)

12

Some Legal Puzzles in the 1996 Statutory Provisions for Universal Telecommunications Services

Warren G. Lavey
Skadden, Arps, Slate, Meagher & Flom (Illinois), Chicago, IL

The Telecommunications Act of 1996 (Pub.L. No. 104-104, The Act) made great strides in stating universal service principles and providing for the development of associated support mechanisms by the Federal Communications Commission (FCC). The previous statutory hooks for universal service policies and practices, found in the Communications Act of 1934, were limited to the broad goal of "nation-wide" availability of communications service (Section 151), and to a far lesser extent the provisions requiring no unreasonable discrimination against any class of persons or locality (Section 202) and "just and reasonable" charges (Section 205). Now, in Section 254 of the Act there are pages of statutory universal service policies, standards, and procedures.

Like much of the Act, many details are left for the FCC to fill in and many important issues are not addressed by the statute. In addition, the 1996 statutory provisions for universal service contain at least three "puzzles" that the FCC, state regulators, and the courts will have to work through. First, how can the FCC assure that federal support mechanisms are "sufficient" given that state support mechanisms and intrastate ratemaking are outside of the FCC's jurisdiction? Second, what is the basis for ratemaking (cost recovery) applicable in determining whether federal support mechanisms are "sufficient"? Third, how will the FCC's support mechanisms be "sufficient" and "predictable" when the telephone industry has more than 1,000 local carriers eligible for universal service support, tens of thousands of areas whose costs and rates must be considered, and diverse operating conditions with outliers?

This chapter examines the incomplete nature of the statutory provisions surrounding these puzzles and offers suggestions for their resolution by the FCC, state regulators, and the courts. Some other issues posed by the universal service statutory provisions are attracting greater attention, such as how to define universal service, what services or revenues should be subject to charges to support universal service, and how to determine the amounts of support payments. (See *Federal–State Joint Board on Universal Service*, 1996.) Nevertheless, these statutory puzzles have the potential to bedevil universal service rules for many years, much like the FCC has encountered in other areas of implementing the Act, such as interconnection. (*Iowa Utilities Board v. FCC*, 1997).

FINGER POINTING—WHO IS NOT SUFFICIENT?

Before passage of the Act, federal regulators had jurisdiction over rates, terms and providers of interstate services, and state regulators had jurisdiction over intrastate services (*Communications Act of 1934*, Section 152(b); *Louisiana Public Service Commission v. FCC*, 1986). Despite the limited jurisdiction of the FCC, federal regulators embraced the policy goal of universal service that focused on the level of local service rates (an intrastate offering). The FCC supported universal service through jurisdictional separations formulae that allocated joint and common costs of local exchange carriers (LECs) between interstate and intrastate services for purposes of interstate ratemaking, as well as interstate recovery of certain costs for connecting and providing local service to low-income subscribers (*Amendment of Part 36 of the Commission's Rules and Establishment of a Joint Board*, 1994/1995; *Federal–State Joint Board on Universal Service*, 1997; *MTS and WATS Market Structure*, 1983). A generous bite of local exchange costs—especially for small, rural companies—recovered by interstate rates reduced the need for high local rates.

The Act reshapes the roles and authority of federal and state regulators in several areas, with the exact contours of the new shapes yet to be clarified by the courts. In the area of universal service, one of the jurisdictional issues that must be addressed is the standard of sufficiency for the federal support mechanisms. The puzzle emerges from four interrelated provisions of Section 254 of the Act.

First, in the statement of principles for the Federal–State Joint Board and the FCC to base policies for the preservation and advancement of universal service, the Act includes the following principle: "There should be specific, predictable and sufficient Federal and State mechanisms to preserve and advance universal service" (Section 254(b)(5)). This principle appears to require, or at least allow the Joint Board and FCC to assume, that state mechanisms exist to preserve and advance universal service (*Federal–State Joint Board on Universal Service*, 1997). But, what can be assumed at the federal level about such state mechanisms?

The statutory language can best be read as applicable to the aggregate of federal and state mechanisms; in total, they must be specific, predictable, and sufficient to preserve and advance universal service. This aggregate standard may mean that each federal or state mechanism must be specific and predictable so that the aggregate of mechanisms is specific and predictable. However, only the aggregate of mechanisms must be sufficient to preserve and advance universal service; it makes no sense to envision duplicate support mechanisms, each of which is independently sufficient to support a common definition of universal service. This statutory principle cannot be viewed as authorizing the FCC to prescribe state mechanisms that are sufficient. This principle describes state mechanisms as distinct from federal mechanisms, which is consistent with the tradition of independent development of federal and state support mechanisms as well as the preservation of state authority under Section 254(f) of the Act (*Access Charge Reform*, 1997).

It follows from applying the sufficiency standard to the aggregate of federal and distinct independent state mechanisms that the federal mechanisms cannot be determined as compliant with the statute without consideration of the state mechanisms. But, does this principle require the federal mechanism to compensate for a particular state's mechanism that provides substantially less support for universal service than other states' mechanisms (the "hypothetically deficient state mechanism")? Put differently, would the federal mechanism comply with this statutory principle if it was sufficient when viewed together with a standard for state mechanisms that the federal mechanism is allowed to assume for each state? These issues are returned to after considering three other statutory provisions regarding sufficient mechanisms.

Second, the statute went beyond the statement of the sufficiency principle in Section 254(b)(5) to describe the standard for the support mechanisms in two other provisions. One of these provisions addresses contributions by telecommunications carriers, and the other addresses recipients and uses of the support.

Section 254(d) on contributions begins with the following requirement imposed on carriers: "Every telecommunications carrier that provides interstate telecommunications services shall contribute, on an equitable and nondiscriminatory basis, to the specific, predictable, and sufficient mechanisms established by the Commission to preserve and advance universal service." This reference to sufficient mechanisms established by the FCC differs from the sufficiency principle in Section 254(b)(5) covering federal and state mechanisms. Does this mean that the federal mechanisms, taken alone regardless of the sufficiency of state mechanisms, must be sufficient to preserve and advance universal service?

The better reading is that this section does not preclude consideration of other factors (including state support mechanisms) in determining whether federal mechanisms are sufficient. On the other hand, this section does not address the issue just discussed as to the federal obligation in light of a hypothetically deficient state mechanism.

Third, and similarly, in addressing the recipients and uses of federal universal service support, Section 254(e) does not address the puzzle flowing from the sufficiency principle. This section provides: "Any such [Federal universal service] support should be explicit and sufficient to achieve the purposes of this section." Aside from the difference in the description of the sufficiency standard—sufficient to preserve and advance universal service versus sufficient to achieve the purposes of this section—this section does not add to the statutory guidance on how to determine the sufficiency of federal mechanisms in light of state jurisdiction.

The fourth relevant statutory provision specifically addresses state mechanisms, but does not appear to solve the puzzle of federal obligations in light of a hypothetically deficient state mechanism. Section 254(f) provides:

> A State may adopt regulations not inconsistent with the Commission's rules to preserve and advance universal service. Every telecommunications carrier that provides intrastate telecommunications services shall contribute, on an equitable and nondiscriminatory basis, in a manner determined by the State to the preservation and advancement of universal service in that State. A State may adopt regulations to provide for additional definitions and standards to preserve and advance universal service within that State only to the extent that such regulations adopt additional specific, predictable, and sufficient mechanisms to support such definitions or standards that do not rely on or burden Federal universal service support mechanisms.

According to Section 254(f), states must allow implementation of the federal mechanisms. For example, if the federal mechanism specifies certain services which are intended to use federal support, state ratemaking cannot require carriers to charge rates that are inconsistent with such intended support. Any reading of the first sentence of Section 254(f) as authorizing the FCC to adopt rules that prescribe, or even impose requirements as to the sufficiency of, state mechanisms appears to be undermined by the second sentence of the same section. The second sentence provides that states determine the manner of contribution for state mechanisms.

As for the third sentence of Section 254(f), states may go beyond the definition of universal service supported by federal mechanisms, but if so states must provide sufficient support for such additions without relying on or burdening federal mechanisms. However, this provision does not deal with any requirement or standard for the sufficiency of state mechanisms in supporting the definition of universal service adopted for purposes of federal mechanisms. The language used here regarding independent, self-sufficient state mechanisms for additions to the federal universal service definition contrasts with the aggregate federal–state approach to the federal definition of universal service in the sufficiency principle of Section 254(b)(5). This language also contrasts with the requirement of sufficient federal mechanisms in Sections 254(d) and (e). Congress never said that federal mechanisms must not rely on or burden state universal service support mechanisms.

After reviewing the puzzle posed by these four statutory sections, how should the sufficiency of federal mechanisms be determined? The best solution is to allow the Joint Board and FCC to develop reasonable assumptions about support from state mechanisms. The reasonableness of these assumptions could be tested against historic levels of state support for universal service, ongoing developments of new support mechanisms by the states, the relative burdens carried by federal mechanisms versus what is assumed to be carried by state mechanisms, and other factors. Courts should require the FCC to develop a record demonstrating the reasonableness of the FCC's assumptions about state mechanisms. The sufficiency of federal mechanisms could be examined in light of such reasonable assumptions about state mechanisms.

The federal assumption would not take the form of requirements for or limitations on state mechanisms. A state could, pursuant to Section 254(f), choose to provide less support for the federally defined universal service than the federal assumptions. But, such a hypothetically deficient state mechanism would not cause the federal mechanism to become insufficient in violation of the statute.

MEASURING THE NEED—WHAT COSTS AND RATES ARE RELEVANT?

The second puzzle derives from the limited federal jurisdiction over LECs' cost recovery and rate designs. Much like the first puzzle dealing with independent state support mechanisms, this second puzzle confronts the FCC with satisfying a standard of sufficiency where major factors are unknown to it and beyond its control.

The rates charged for any telecommunications service depend on the costs to be recovered by that service as well as the rate design for that service. Much of the costs of LECs are joint and common for multiple services, requiring allocations to specific services. Federal and state regulators use a wide variety of methodologies for such allocations, and generally cost allocations across intrastate services are beyond the FCC's jurisdiction. Also, there are substantial differences in state regulators' authorized rates of return and accounting practices for LECs, which affect costs and rates for intrastate services. Moreover, costs can be defined based on historic (embedded) costs, forward-looking costs, or proxy measures of costs (such as price caps).

Regarding rate design, a service such as local calling has a wide variety of rate elements that differ from state to state and from carrier to carrier—flat monthly charges, measured usage charges (based on distance called, time of day, and duration), installation charges, geographic and population calling scope, business versus residential line, digital versus analog service, and so forth.

State regulatory decisions regarding the costs recovered by an intrastate service and rate design for an intrastate service are important factors in determining whether federal support mechanisms are sufficient. In addition to the sufficiency principle of Section 254(b)(5) quoted in the preceding section, Sections 254(b)(1) and (3) of the statute provide the following two universal service principles which are tied to rate levels:

> Quality services should be available at just, reasonable, and affordable rates.
>
> Consumers in all regions of the Nation, including low-income consumers and those in rural, insular, and high cost areas, should have access to telecommunications and information services, including interexchange services and advanced telecommunications and information services, that are reasonably comparable to those services provided in urban areas and that are available at rates that are reasonably comparable to rates charged for similar services in urban areas.

Despite the Act's focus on rates (including rates for intrastate services within the jurisdiction of state regulators), the Act provides little direction on the development of rate benchmarks for the federal mechanisms. The only statutory provision is Section 254(k), which prohibits cross subsidies that would inflate the need for universal service support: "The Commission, with respect for interstate services, and the States, with respect to intrastate services, shall establish any necessary cost allocation rules, accounting safeguards, and guidelines to ensure that services included in the definition of universal service bear no more than a reasonable share of the joint and common costs of facilities used to provide those services." In addition to other sections of the statute, such as Section 152(b), this provision clearly does not authorize the FCC to preempt, or even to impose limitations on, state ratemaking for intrastate services' rates or cost allocations. The only comfort that the FCC can derive from this section is that federal mechanisms should not be determined to be insufficient based on the existence of certain high intrastate service rates if such rates bear more than a reasonable share of joint and common costs. How can the sufficiency of federal mechanisms in making rates "affordable" and "comparable" be determined when ratemaking practices vary greatly from state to state and from carrier to carrier? The best solution to this second statutory puzzle is like what was described in the preceding section.

The Joint Board and the FCC should develop reasonable assumptions about intrastate services' rates by area. The reasonableness of these assumptions should be supported by a record reflecting actual rates and state ratemaking practices. For example, if states typically base rates on recovery of LECs' actual historic costs rather than forward-looking costs, then the federal mechanism must be sufficient in light of such typical cost recovery.

The federal assumptions would not constrain state regulators' authority over cost recovery and rate design for intrastate services. Yet, these assumptions would allow the FCC to show the sufficiency of mechanisms in light of diverse state ratemaking practices that are beyond the FCC's control.

PREDICTABLE AND SUFFICIENT—FOR ALL RECIPIENTS?

The third puzzle in the Act's provisions on universal service revolves around the dual standards of predictable as well as sufficient support mechanisms. These standards appear together in the statement of universal service principles in Section 254(b)(5) and in the Section 254(d) provision on contributions by telecommunications carriers. As discussed here, it may be relevant that the Section 254(e) provision on recipients and uses of universal service support requires such support to be explicit and sufficient, but does not employ the language of specific, predictable, and sufficient appearing in Sections 254(b)(5) and (d).

Recognition of this puzzle of pursuing predictability as well as sufficiency depends on two relevant facts. First, there are more than 1,000 LECs eligible for support, tens of thousands of areas whose costs and rates must be considered, and diverse operating conditions with outliers. Second, the FCC has often dealt with outliers such as small, rural LECs on an exception basis. For example, the FCC allowed certain carriers to charge interstate access rates based on their individual historic costs instead of forecasted costs or price caps, or to change the definition of study areas for high-cost funds (*Alltel v. FCC*, 1988; *Access Charge Reform*, 1997; *National Rural Telecom Association v. FCC*, 1993). Yet, any ability to apply for exception treatment involves inherent delays and uncertainties.

The diversity of operating conditions means that it would be highly demanding to require that the federal support mechanisms be both predictable and sufficient for each area, including the outliers. The standard of both predictability and sufficiency becomes especially onerous if predictability means that the FCC cannot use the individualized, actual costs of outliers or case-by-case exceptions to support rules. Such flexibility in administrative treatment may be necessary for the federal mechanism to provide sufficient support for outliers, but it sacrifices some level of predictability for the recipients as well as the contributors.

The answer to this puzzle lies in the fact that the outliers among eligible LECs comprise a small percentage (by one measure under 8%) of the industry measured in telephone access lines served (*Access Charge Reform*, 1997).

Returning to the statutory language, Section 254(d) requires that carriers contribute to predictable and sufficient mechanisms. In this context, the standard of predictability shall be applied to the amounts of contributions charged to individual carriers. As long as the process of exception treatment would likely affect an individual carrier's contributions by only a few percent, the mechanism should be determined to comply with the predictability requirement of this section. Accordingly, the predictability standard here does not preclude an exception process within a federal support mechanism, but does require that the applicability of such exception process be limited to small amount of the federal support.

This limitation could be achieved by allowing only areas comprising a small percentage of the telephone access lines served to apply for exception treatment and restricting the total amount of support that would be provided through exception treatment. Put differently, the support mechanism must provide sufficient support for the vast majority of areas without exception treatment.

As for Section 254(e) on recipients and uses of support, no requirement of predictability is provided here, only the standards of specificity and sufficiency. An exception process could have a substantial impact on the amount of support received for an area. Also, the uncertainties as to timing and amount from a case-by-case approach may make this process less predictable for areas that use it than for other areas. Yet, all that Section 254(e) appears to require specificity for are the procedures and evaluation criteria of any exception process as well as enough attention to the needs of outliers to assure sufficient support for each area.

Of course, a level of predictability for each recipient is also required by the general requirement that the FCC's decisions cannot be arbitrary or capricious (Administrative Procedure Act, Sec. 706(2)(A)). The FCC must develop record evidence for the amount of support provided to any area, and describe a reasoned basis for providing different amounts of support to different areas.

In summary, there is some tension between the standard of predictable support and the standard of sufficient support in light of the diversity of areas. The solution to this puzzle lies in not interpreting the predictability standard as precluding a case-by-case approach. However, such departure from explicit formulae should be applicable only to certain areas where such exception treatment is necessary to obtain sufficient support for a limited number of areas with a small impact on the total contributions required from carriers.

CONCLUSION

No sweeping statutory change is perfectly specified, and the universal service sections of the Act should not be found wanting because of the existence of certain puzzles. The Joint Board, FCC, state regulators, and courts—as well as carriers, consumer representatives, and other interested parties—should all attempt to find the best solutions to any puzzles within the framework of what Congress has given us.

For the three puzzles explored in this chapter, there appear to be solutions that are consistent with the authority and reasonable capabilities of the FCC as well as with the language and purposes of the universal service sections of the Act.

REFERENCES

Access Charge Reform, 12 FCC Rcd 15982, 15988, 16126-28 (1997)

Administrative Procedure Act, 5 U.S.C.A. Sec. 706 *et seq.* (West 1996 & Supp. 1997)

Alltel v. FCC, 838 F.2d 551 (D.C. Cir. 1988)

Amendment of Part 36 of the Commission's Rules and Establishment of a Joint Board, 9 FCC Rcd 7404 (1994), 10 FCC Rcd 12309 (1995)

Communications Act of 1934, 47 U.S.C.A. Sec. 151 *et seq.* (West 1991 & Supp. 1995).

Federal–State Joint Board on Universal Service, CC Docket No. 96-45, 12 FCC Rcd 87 (1996), 12 FCC Rcd 877 6, 9192–97 (1997)

Iowa Utilities Board. v. FCC, 109 F. 3d 418 (8th Cir. 1996), 120 F.3d 753 (8th Circ. 1997), *cert. granted*, 118 S. Ct. 879 (1998)

Louisiana Public Service Commission v. FCC, 476 U.S. 355 (1986)

MTS and WATS Market Structure, 93 FCC 2d 241 (1983)

National Rural Telecom Association v. FCC, 988 F.2d 174 (D.C. Cir. 1993)

Telecommunications Act of 1996, Pub. L. No. 104-104, 47 U.S.C.A. Sec. 151 *et seq.* (West Supp. 1997)

13

Universal Service and the National Information Infrastructure (NII): Making the Grade on the Information Superhighway

James McConnaughey
National Telecommunications and Information Administration

Policymakers in this country have long embraced universal service as a broad and purposeful goal. Throughout this century, the number and proportion of U.S. households with telephones have increased steadily as "plain old telephone service" (POTS) has come within the economic reach of most Americans (see AT&T, 1982; Belinfante, 1998; FCC, 1997; Lande, 1993). Developed within the structure of franchised local exchange monopolies, the policy flourished through the workings of a system that featured wide-ranging subsidies and noncost-based telephone rates (see, e.g., Weinhaus et al., 1992). For much of this century, this arrangement has stood as the legacy of the Bell System's Theodore Vail and the Communications Act of 1934 (see, e.g., Bolter et al., 1990).

In recognition of a profoundly new era based on telecommunications and computers, the Clinton Administration launched the National Information Infrastructure (NII) initiative. On September 15, 1993, the Administration's Information Infrastructure Task Force (IITF, 1993) issued the NII *Agenda for Action*, a policy blueprint that sets forth nine major principles and goals for government activity. One of those principles is to "extend the 'universal service concept' to ensure that information resources are available to all at affordable prices" (p. 6). More specifically, the *Agenda for Action* directed the Commerce Department's National Telecommunications and Information Administration (NTIA) to conduct universal service hearings around the country. The Administration's vision statement also instructed the Commerce-chaired IITF to work with the NII Advisory Council, as well as state regulatory commissions, in determining a universal service concept for the 21st century. This chapter discusses: (a) the results achieved by universal service and related policies in this country, and (b) how federal actions are helping to expand this traditional concept from dial tone for households to basic information technology access for all Americans.

TRADITIONAL UNIVERSAL SERVICE

Progress toward the universal service goal is generally measured by the metric "telephone penetration," which in its most widely used form represents the percentage of inhabitants or households that have a telephone on premises. Using this basis, the United States has achieved some important successes. For example, comparing main telephone lines per 100 inhabitants using 1996 data, the United States (64.0) leads Canada (60.2), France (56.4), Germany (53.8), the United Kingdom (52.8), Japan (48.9), and Italy (44.0) among the group of seven countries, and trails only Sweden (68.2) and Switzerland (64.0) among all major industrialized nations (International Telecommunication Union [ITU], 1998). And the U.S. growth in main lines per 100 (17.3%) has been more robust than that of Sweden (-.1%) and exceeded only by Germany (34.0%) and the U.K. (19.4%) among "G-7" nations during the 1990s (ITU, 1998). Since the breakup of AT&T's Bell System in 1984, the average annual percentage of U.S. households without telephones has declined by about one fourth (-24.3%), whereas the total number of households in this country increased by almost one fifth (19.5%) to a total of 102 million (Belinfante, 1998).

A closer examination, however, reveals that important groups or geographic areas exhibit telephone penetration rates that significantly lag behind the national average. For example, on the San Carlos Apache Reservation in Arizona, only 28% of households currently subscribe to telephone service (USDA, 1997). In determining that 94.8% of U.S. households had at least one telephone, the 1990 U.S. decennial census also found that those less likely to have a telephone included the following types of households (HHs): American Indians (76.8% penetration rate; 47.0% on reservations); those with incomes under $5,000 (77.0%); single-parent homes headed by a female (84.3%); HH head under 25 years of age (84.7%); mobile home occupants (86.9%); households numbering seven or more persons (88.1%); renters (89.3%); those living in nonmetropolitan areas (91.3%); and female HH head (93.5%; U.S. Bureau of the Census, 1994b).

Subsequent studies by NTIA and the Federal Communications Commission (FCC), using special Census survey data, add to this profile. For example, NTIA's *Falling through the net* (McConnaughey, Nila, & Sloan, 1995) analysis highlighted that the phoneless are particularly concentrated in rural areas and central cities; in fact, central cities in the northeast registered the lowest penetration rates, followed closely by central cities and rural areas in the south. The less educated also are increasingly likely to be without telephones. The Commission found that the southwest and south generally lag behind the rest of the United States, with the nation's lowest 1997 average penetration rates in New Mexico (88.1%), Arkansas (89.8%), and Mississippi (89.2%), well behind the national average of 93.9%. Both the FCC and NTIA documented that Black and Hispanic households exhibit relatively low penetration as well (Belinfante, 1998; see also Schement, Belinfante, & Povich, 1994; Wilhelm, 1996).

Why does phonelessness occur? Much work needs to be done, but some important research to date indicates several key factors at play. Some people may eschew having a telephone because they simply do not desire it. For example, in research done by the University of Texas (Horrigan & Rhodes, 1995), 5% of the respondents in a survey of 172 Texans without telephones stated that they did not want or need telephone service. Based on a study of a New Jersey city by two Rutgers University researchers (Mueller & Schement, 1995), many inner-city households apparently prefer cable television service to telephone service. Reasons include: (a) the perceived uncontrollable costs of phone service; (b) telephones could permit undesirable interactions relating to drugs and crime; and (c) phones would allow calls pertaining to bill collection or other "threatening" matters. In contrast, cable is seen as inexpensive, satisfying entertainment that can keep children off dangerous streets and represents a "visible sign of well-being" (Mueller & Schement, 1995). Nationwide, the incidence of those who will not receive telephone service regardless of its availability is likely to be small—less than 1% of total households (Schement et al., 1994).

Others may not have a phone because of language barriers. A witness testified during NTIA field hearings held in South Central Los Angeles, on February 16, 1994, that, based on a 1992 survey, more than 40% of Chinese, Korean, and Latino consumers in San Francisco, Los Angeles, New York, Chicago, and Brownsville, Texas, experienced difficulties in beginning the process of getting a phone installed (NTIA, 1994a).

Affordability of telephone service appears to be a very significant factor. Interestingly, recent studies (see C&P Telephone Company, 1993; Field Research Corporation, 1993; Horrigan & Rhodes, 1995; Mueller & Schement, 1995; Southwestern Bell Telephone, 1986) point to an inability to control long-distance calling as a key determinant that, overall, is apparently more important than the current cost of local service, including basic monthly service rates and installation charges. Past outstanding bills, including the cost of reinstallation, also seem to pose a substantial barrier to phone service for many of today's phoneless (see, *e.g.*, Horrigan & Rhodes, 1995).

UNIVERSAL SERVICE IN THE INFORMATION AGE

The information marketplace is substantial and growing. NTIA estimates that information and telecommunications equipment and services generate more than three quarters of a $1 trillion in revenue annually in the

United States and may comprise 17% of this country's gross domestic product by the early 21st century (Irving, 1997b). With private investment in such infrastructure as the catalyst, the resulting technological innovations and wider access will boost U.S. global competitiveness through higher productivity and new products and services. Expected economic benefits include the creation of high-skilled and better paying jobs, economic growth, and an increased standard of living for Americans. Important social benefits would arise from greater educational opportunities such as distance learning, expanded medical applications through telemedicine linkage of remote locations with big-city hospitals, and community empowerment (see Brown, 1995; IITF, 1993).

Clearly, information is increasingly impacting U.S. lives, and many will be favorably affected. However, there is a serious concern about how others will fare—the so-called information poor. Indeed, those already disadvantaged persons who do not participate in the information revolution will fall even further behind the mainstream in terms of their economic well-being. The Clinton Administration seeks to ensure that such widening gaps between information "haves" and "have nots" do not occur (IITF, 1993).

A fundamental goal of this Administration is to expand the universal service concept to ensure that information resources are available to all at affordable prices (IITF, 1993). This applies both to households that are not currently connected to the NII, and to public institutions that are traditionally providers of information access to the general public. Both President Bill Clinton (1994, 1996a, 1997, 1998a) and Vice President Al Gore (1994, 1998a) believe that public policymakers should work with the private sector to connect every classroom, library, hospital, and clinic in America to the NII by the year 2000.

Ideally, all households that desire access to the NII should be able to achieve it, particularly those persons who would otherwise be information poor. This should remain a long-term national strategy. To advance the goal of universal service in the interim, transitional arrangements need to be made utilizing appropriate institutions as public safety nets.

Where do households and institutions such as public schools and libraries stand in terms of information access? As measured by such gauges as penetration of telephones, personal computers (PCs) and modems, and online access, progress has been made in recent years but much more is needed. With respect to households and their access to phones, the previous section discussed the impressive aggregate numbers but concurrent deficiencies within certain groups and geographic areas. Not unexpectedly, the PC and modem penetration levels are much lower: among all U.S. households, 36.6% possess a PC and 26.3% have modems, up from 25.5% and 11.0% in 1994. (U.S. Bureau of the Census, 1994a, 1997). Similar to telephones, aggregate figures mask the fact that some groups and areas lag substantially behind others. Thus, information have nots are disproportionately found in rural locales and central cities (McConnaughey et al., 1995).

More specifically, the rural poor, followed by low-income households in central cities, have the lowest PC and modem penetration rates (McConnaughey et al., 1995). Senior citizens and the youngest household heads (25 years or younger), as well as Blacks and Hispanics in central cities and particularly rural areas, are also deficient in this regard (McConnaughey et al., 1995). For a given level of education, central city households registered the lowest PC penetration, whereas in rural locations generally the relative incidence of modems trailed all others (McConnaughey et al., 1995). As was the case with telephones, the Northeast's central cities claimed the lowest PC penetration, whereas modem possession ranked lowest among PC users in the rural west, midwest, and south (McConnaughey et al., 1995). Recent evidence has been collected (Katz & Aspden, 1997) that household Internet usage has been characterized by a large dropout rate, particularly among the young, less affluent, and the less educated. Presumably, the high cost relative to one's income level, perceived lack of user friendliness, and the absence of an online "killer" application have hampered the spread of PCs and electronic access, but research is needed to determine more definitively the causes.

Public institutions such as schools and libraries have exhibited a pattern of penetration similar to households: progress over time but room for significant improvement. At public schools, telephones typically are located in and around administrative offices but generally reach only about 12% of the classrooms (National Education Association and Princeton Survey Research Associates, 1993). Although the percentage of students per personal computer has improved tenfold over the past decade, online access to the Internet or other wide area networks appears to be suboptimal (Quality Education Data, Inc., 1995). Overall, 78% of public schools and only 27% of all instructional rooms (classrooms, labs, and media centers) are connected to the Internet (National Center for Education Statistics [NCES], 1998). Some important demographic-based distinctions do exist. For example, schools with low minority enrollment (less than 6%) feature Internet access in 37% of their instructional rooms, compared to 13% in predominantly minority (50% or more) schools (NCES, 1997). Less pronounced is the access differential between instructional rooms in rural (30%) and city schools (20%; NCES, 1998). Concerning Internet capabilities, teachers have greater access to the World Wide Web (WWW; 94%) and particularly email (88%) than do students (74 and 35%, respectively) at public schools (NCES, 1997).

Virtually 100% of public libraries appear to have at least one telephone on premise, albeit almost two thirds (63%) have five or less (National Commission on Libraries and Information Science [NCLIS], 1994). Concerning urban libraries, 85% are characterized by 10 or more phone lines, whereas 79% of rural libraries have one to five lines. Of greater concern to policymakers is that more than one third (37%) of public libraries do not own computers (NCLIS, 1994). Although 63% of public libraries possess one or more computers, only a little more than 6% exceed 10 in number (NCLIS, 1994). Regarding computers with external communications capability, 72% of all public libraries are connected to the Internet (NCLIS, 1997b). For those libraries that do have Internet access, only 60% provide public access terminals (NCLIS, 1997b). Substantially higher proportion of urban libraries are hooked up to the Internet than their rural counterparts serving populations of less than 5,000 (66%), (NCLIS, 1997b). Although the highest percentage of public access to graphical WWW is found in rural areas (32% of connected libraries), these same areas also possess the highest proportion of main/central libraries with public access to those interfaces (17.2%; NCLIS, 1997b).

These institutions have inadequate communications capabilities for several reasons. With respect to telephones, budgetary factors and (in the case of schools) teacher attitudes appear to be important (Office of Technology Assessment, 1995). Multiple barriers prevent schools and libraries from adequately accessing the NII (e.g., the Internet). The most important reasons that schools cite as to why they do not have or use advanced communications include limited funding, too few access points in the building, lack of or poor equipment, and not easily accessible telecommunications equipment or "links" (National Center for Education Statistics, 1997a).

For libraries, important factors center on cost considerations, the availability of state money, library staff's time and technical expertise, and federal money. For urban libraries, system/server costs, communications costs, and staff time are most critical. Internet access by rural libraries also hinges most importantly on these two types of costs, followed by state money, software costs, staff time, and in-house technical expertise (NCLIS, 1997a).

ADMINISTRATION ACTIVITIES

The Administration has undertaken a number of activities under the aegis of the IITF and the NII initiative that are designed to preserve and advance universal service. Simply stated, these activities collectively seek to avoid creating a society of information haves and have nots.

Universal Service Hearings

As the chair of the IITF Telecommunications Policy Committee (TPC) and its Working Group on Universal Service, NTIA—led by its Administrator, Larry Irving—cosponsored with several state governments five universal service field hearings in different parts of the country during 1993 and 1994. Featuring 230 witnesses and 1,400 pages of transcript, the hearings sought to determine how the concept of universal service can be made more consistent with the needs of Americans today and in the 21st century. In September 1994, NTIA released *America Speaks Out*, a report that identifies major themes of these hearings, including most fundamentally that government has an important role to play in ensuring that the goals of universal service are accomplished in a multiprovider environment (NTIA, 1994b).

Notice of Inquiry

In September, 1994, NTIA also issued a Notice of Inquiry (NOI) on universal service and open access as part of a comprehensive review of universal service and related issues in communications (NTIA, 1994a). The NOI reflects the current debate regarding whether traditional notions of universal service—the widespread availability of "basic" telephone service—are adequate to meet the needs of the American people. Information has emerged as a vital economic resource and source of individual empowerment. Because technological changes are making telecommunications networks the highways of the information age, businesses and individuals must be able to access and distribute essential information resources. In addition, the emergence and expansion of competition is forcing the United States to reexamine traditional U.S. support policies. Simultaneously, technological changes and the emergence of the Information Age are prompting a fresh look at the way universal service is defined.

Some 98 private or public sector entities, as well as individuals, submitted comments on the NOI. NTIA is drawing on the record of this proceeding in developing analyses relating to several facets of the universal service issue.

Electronic Conference

In November 1994, NTIA and the TPC held the first-ever conference in which the government has sought public input entirely through electronic networks. The Administration produced this electronic or "virtual" conference to: (a) garner opinions and views on universal service that may shape the legislative and regulatory debate; (b) demonstrate how networking technology can broaden participation in the development of government policies, specifically, universal service policy; (c) illustrate the potential for using the NII to create an electronic commons; and (d) create a network of individuals and institutions that will continue the dialog started by the conference (Irving, 1994). More than 900 inputs from private locations and 78 public access points in the United States produced comments on such topics as redefining universal service, availability/affordability, and subscribership (including access for the disabled community; NTIA, 1994c).

Federal–State-Local Telecom Summit

On January 9, 1995, the Administration, led by Vice President Gore and the Commerce Department (with cosponsorship by the Annenberg Washington Program), conducted an intergovernmental summit of telecommunications policymakers in Washington, DC. Among the principles agreed to as a framework for discussion, two directly pertained to universal service. The first emphasized the importance of affordable, just, and reasonable prices for access to basic telecommunications, whereas the second stressed that telecommunications providers must contribute to the maintenance of universal service on an equitable and competitively neutral basis (Gore, 1995). One year later, the new Telecommunications Act of 1996 embraced both concepts.

Household Penetration Study

Using November 1994 Current Population Survey data from the Census Bureau, NTIA developed a landmark subscribership study of U.S. households called *Falling through the net* that the agency released in July 1995 (McConnaughey et al., 1995). In the case of telephones, personal computers, and online access, have nots are disproportionately represented in certain geographic areas, such as rural designations and central cities in particular regions (e.g., northeast, south). As discussed, those who are most likely to be information disadvantaged—yet enthusiastic users, once online—are found among the poor, minorities, young, and those who are less educated.

In April 1998, Vice President Gore directed NTIA to conduct a study of the "digital divide" of Internet and computer use among races and income levels (Gore, 1998a). Soon thereafter, Commerce Department Assistant Secretary Irving announced that "trendline" (1984–1997) analyses and a new expanded "Falling through the net" study would be undertaken during 1998–1999 (Irving, 1998).

Access for Schools, Libraries, and Health Care Providers

During mid-1995, NTIA issued *Connecting the Nation: Classrooms, Libraries, and Health Care Organizations in the Information Age*, a review of the status of NII connectivity at public institutions (Gonzalez, 1995). The report highlighted that, as of 1994, Internet access numbered 35% of K–12 public schools, 21% of public libraries, and 22% of hospitals (Gonzalez, 1995). Overall, *Connecting the Nation* concluded that "much work" needs to be done before the Administration's goal of connecting all classrooms, libraries, and health care organizations can be met (Gonzalez, 1995). As discussed below, important progress has been achieved since that time.

In February 1996, the Administration at the highest levels began promoting an educational technology initiative, called the "Technology Literacy Challenge," whereby $2 billion in trust funds would be provided to state and local governments—which would be encouraged to seek private matching funds—for permitting students and teachers to access and successfully use the NII (see U.S. Department of Education, 1996). The "four pillars" of the Challenge initiative include:

1. Modern computers and learning devices will be accessible to every student.
2. Classrooms will be connected to one another and to the outside world.
3. Educational software will be an integral part of the curriculum.
4. Teachers will be ready to use and teach with technology. (See White House, 1997.)

Other activities designed to assist schools are also underway. Federal agencies and the Lazarus Foundation are working to implement the President's Executive Order directing that surplus government computers should go first to schools, with priority for schools in "empowerment zones" (White House, 1996). In 1995, the private sector with a general endorsement by the Administration established the "TECH CORPS," which marshals technical volunteers to train teachers in use of the NII. Specifically, the mission of TECH CORPS is to bring and support volunteers from the technology community to advise and assist schools with respect to new educational technologies. Assistance focuses on local planning, technical support and advice, staff training, mentoring, and classroom interactions. Funded principally through corporate sponsorships, the nonprofit organization currently operates 42 state chapters and one in the District of Columbia (TECH CORPS, 1998). In May 1996, President Clinton announced the creation of 21st-Century Teachers, a nationwide voluntary program that encourages 100,000 teachers to mentor their colleagues to effectively use technology for teaching and learning. The initiative benefits from the involvement of leading education groups and TECH CORPS, which have asked the nonprofit McGuffey Project to spearhead the program (21st-Century Teachers Network, 1997).

Reminiscent of the community barn-raising activities of a more agrarian economy, NetDay was launched as a voluntary program to provide internal connections needed by schools to access the Internet. Inaugurated in California in March 1996, the initiative brought together approximately 20,000 volunteers and 2,000 businesses to connect classrooms in 4,000 schools to the Internet (on one day [March 9]). On June 29, 1996, Vice President Gore, NTIA, the Department of Education, and others participated in the "NetDay 96 How To" conference in Washington, DC (Gore, 1996b). More than 40 states held NetDays in 1996, many of them on a national NetDay held on October 25 (Gore, 1997a). On February 7, 1997, Assistant Secretary of Commerce and NTIA Administrator Larry Irving encouraged business leaders to develop public–private partnerships with schools similar to the Adopt-A-Highway program (Irving, 1997a). Another national NetDay took place on April 19, 1997, and now virtually every state has participated in the program (NetDay 2000, 1998).

Assisting Rural Access

In October 1995, NTIA released its *Survey of Rural Information Infrastructure Technologies*, which describes rural information applications that make use of voice, computer, and video telecommunications services; and delineates various wireline and wireless technologies that are or could be used to deliver these services to rural areas (NTIA, 1995). The report concludes that: (a) there are a number of technologies suitable for providing voice grade services in rural areas; (b) it is technically feasible to provide advanced computer and video capabilities in many such areas; and (c) there is currently no technology, however, that can economically provide broadband capabilities to the "most isolated farms, ranches, and homes" (NTIA, 1995).

Infrastructure Grants or Loans for Non-Profit Institutions and Carriers

Some Administration activities related to the NII initiative, such as grants, serve to complement—but are not a part of—the universal service system. Conceptually, they have different objectives: grants made, for example, under NTIA's new Telecommunications and Information Infrastructure Assistance Program (TIIAP) target myriad types of projects that serve as innovative models for other entities to emulate, whereas the new universal service system is designed to provide funding for discounted purchases of telecommunications services, Internet access, or internal connections by eligible schools, libraries, or health care providers. NTIA provides matching grants to assist states, localities, universities and schools, hospitals, and other public information and critical service providers to purchase equipment and undertake planning related to infrastructure development. Since its inception in 1994, TIIAP has generated tremendous interest: more than 4,600 applications have been received and 332 grants have been awarded in 50 states, the District of Columbia, and several territories. These awards of some $100 million in federal monies were supplemented by more than $150 million in nonfederal funds. A significant portion of the funding went to rural regions or states. TIIAP projects serve as models that can be replicated in similar communities across the country, thereby extending their benefits beyond the communities in which they take place (TIIAP, 1998).

Rural Utilities Service (RUS), an agency within the Department of Agriculture, is a rural development agency that assists the private sector in developing, planning, and financing the construction of telecommunications infrastructure in rural America. RUS financing provides loans and some grant monies for carriers to build new networks and modernize existing ones, extend service to unserved areas, and provide facilities for economic development, distance learning, telemedicine applications, and Internet access. In 1995, the agency loaned $585 million to telephone companies that used the monies to provide initial telecommunications services to 75,000 families, install 8,000 miles of fiber-optic facilities, and purchase nearly 200 new digital switches (RUS, 1996).

As part of the Technology Literacy Challenge initiative, the Department of Education provides two types of grants. One, the Technology Literacy Challenge Fund, provides formula grants to states to assist them in implementing statewide technology plans for schools. Technology Innovation Challenge grants are designed to complement Literacy grants by generating new learning applications and best practices that may be adapted successfully across the nation. Both are intended to be financial means to help the nation meet the four pillars goals (U.S. Department of Education, 1996).

National Subscribership Goal

NTIA recommends that the FCC set a national subscribership goal for the year 2000 (Brown, 1996; NTIA, 1996a). Specifically, the goal should ensure that the telephone penetration level for all segments of society is equal to or above the national average that existed as of November 1996. NTIA proposes that the Commission and the states work together to increase subscribership, for example, by jointly exploring changes to the Link-Up program (which supports reduced-cost installation of telephone service) to better serve the needs of low-income, highly mobile users. In addition, based on an empirical study of disconnection and its effect on subscribership, the agency strongly urges that states bar local telephone carriers from disconnecting local telephone service for nonpayment of interstate long-distance charges. As part of making access more readily available to advanced telecommunications and information services, particularly during the interim period before universal household connectivity is achieved, NTIA believes that schools, libraries, and other community access centers should be expeditiously hooked up to the NII (NTIA, 1996a).

The E-Rate Proposal

At the aforementioned June 29 NetDay 96 conference, Vice President Gore announced his support for a proposal put forth by U.S. Education Secretary Richard Riley, Representative Edward J. Markey (D-MA), Senator Byron Dorgan (D-ND), and others for free education or "E-rates" for a package of basic telecommunications services to schools and libraries. In addition, a discounted rate for access to other services would be established. Vice President Gore stated that the E-rate plan would be developed in such a way that: (a) services made available under the plan meet recognized educational objectives, including the need for adequate bandwidth; (b) all competitors have access to the universal service fund subsidies to serve schools and libraries; (c) the subsidies are technology-neutral; and (d) there is a dynamic and open process for defining the services and packages using advice and counsel from all groups that have a stake in the future of the U.S. educational system (see Gore, 1996b).

Subsequently, the Secretaries of Agriculture, Commerce, and Education urged the FCC to "support a guarantee of universal access to advanced telecommunications and information service for every public, private, and parochial school and library in America" (Cabinet, 1996, p. 1). NTIA's companion filing on behalf of the Administration elaborated on the Vice President's E-rate principles, arguing for a system of market-based incentives and universally available telecommunications and information services. Features would include competitive bidding intended to assure a best value combination of price and functionality; tiered discounts tied to economic disadvantage; a procurement process that integrates purchasing decisions, technology plans, and curricula; and regular triennial reviews by the Joint Board and the FCC (NTIA, 1996b). Complements to the E-rate approach would be aggregation of demand through buying coalitions, donations of new or surplus equipment or software, and volunteer activities such as NetDay, training, and technical assistance (NTIA, 1996b).

As the implementation process has proceeded, the Administration has continued its strong support for this program of assisting eligible schools, libraries, and health care providers to participate in the Information Age (Gore, 1998b; Clinton, 1998b).

THE NEW RULES OF THE ROAD

Legislative

In 1996, Congress, working with the Administration and a variety of public and private sector interests, adopted for the first time since 1934 comprehensive legislation that establishes a policymaking framework for the emerging information age. On February 1, 1996, the House and Senate overwhelmingly adopted telecom reform legislation. The final version of the bill (S. 652) was approved in the House 414-16, and 91-5 in the Senate. On February 8, President Clinton signed into law the *Telecommunications Act of 1996* (Pub. L. No. 104-104, 110 Stat. 56, codified at 47 U.S.C. § 151 et seq.). The President and the Vice President commended the tremendous bi-partisan effort that led to passage of the bill (Clinton, 1996c; Gore, 1996b).

Congress made history when it codified universal service as an explicit nationwide policy. Section 254 of the Act defines *universal service* as an "evolving level of telecommunications services that the [FCC] shall establish ... taking into account advances in telecommunications and information technologies and services" (§ 254(c)(1)). The Act directed the Commission to institute a Federal–State Joint Board to develop recommendations on defining and funding universal service, and enumerates several principles (such as nature of access, service quality, and affordable rates) to guide the deliberations. The Joint Board would submit its recommendations to the FCC by November 8, 1996, and the Commission in turn would issue its final rules by no later than May 8, 1997 (§ 254(a)(1),(2)).

In developing a list of the telecommunications services that would be supported by federal funding mechanisms, the Act requires policymakers to consider the extent to which services are: (a) essential to education, public health, or public safety; (b) subscribed to by a substantial majority of residential customers through a market process; (c) being deployed in public telecom networks by carriers; and (d) consistent with the public interest (§ 251(c)(1)).

As a general proposition, the 1996 Telecommunications Act mandates that communications be made available to all people of the United States without discrimination on the basis of race, color, religion, national origin, or sex (§ 104). In addition, it seeks through the so-called Snowe-Rockefeller-Exon-Kerrey provision (named after its Senate sponsors) to ensure that schools and libraries (and rural health care providers) become connected to the NII through preferential rates for universal services as defined by the FCC (§ 254(h)). Moreover, the Act empowers the FCC to designate "special [universal] services" for these institutions (§ 254(c)(3)). Other targeted recipients of support are low-income consumers and those in rural and high cost areas (§ 254(b)(3)). Regarding those that will pay into the system, every telecom carrier that provides interstate and intrastate telecom services shall contribute, on an equitable and nondiscriminatory basis, to the specific, predictable, and sufficient mechanisms established by the FCC and the states to preserve and advance universal service (§ 254(d),(f)).

Regulatory

Charged with implementing these broad directives, regulators have expended enormous energy to provide detailed guidelines and rules for the new universal service policy. On May 8, 1997, the FCC issued a landmark decision regarding universal service—the availability and affordability of telecommunications service for all Americans (Federal–State Joint Board on Universal Service, *Report and Order*, CC Docket No. 96-45, FCC 97-157, rel. May 8, 1997). As directed by the Act, a special Universal Service Joint Board comprised of federal and state regulators and a consumer advocate developed specific recommendations for the FCC with respect to the definition of services that are supported by a federal universal service funding mechanism. Relying heavily on the Joint Board's November 8, 1996, *Recommended Decision*, the FCC released a *Report and Order* that undertakes to modernize universal service policy in an increasingly competitive marketplace and to fundamentally expand its applicability.

Looking to the Act for guidance, the FCC issued rules based on the following goals:

1. All of the universal service objectives established by the Act must be implemented, including those for low-income individuals; consumers in rural, insular, and high cost areas; as well as for schools, libraries, and rural health care providers.
2. Rates for basic service must be maintained at affordable levels for residences.
3. Affordable basic phone service must continue to be available to all users with the help of a universal service fund that will subsidize phone service for those who qualify.
4. The benefits of competition in the telecommunications arena must be brought to as many consumers as possible.

The Commission specifically defined services for households that will be supported by the fund. These include:

1. Access to a telephone network with the ability to place and receive calls.
2. Access to touch-tone capability.
3. Single-party service.
4. Access to emergency systems including, where available, 911 and Enhanced 911.
5. Access to operator services.
6. Access to interexchange services.
7. Access to directory assistance.
8. (For those low-income users who qualify) limited long-distance calling.

The Commission also issued its findings and conclusions with respect to support for rural and high-cost areas, low-income households, schools and libraries, and health care providers (especially rural). Each of these is discussed here in summary fashion.

Rural and High-Cost Areas. Traditionally, universal service has been achieved primarily through implicit subsidies designed to shift costs from rural to urban areas, residential to business customers, and local to long-distance service. Some explicit support mechanisms have been employed in high-cost areas and for low-income users for a number of years. Through changes in universal service and interstate access charges, the FCC will convert existing federal support to an explicit, competitively neutral, and sustainable mechanism as of July 1, 1999 for nonrural carriers. It is intended that costs will be developed through models using forward-looking methodologies. A rural task force appointed by the Universal Service Joint Board will examine the feasibility of developing forward-looking costs for rural carriers. It is anticipated that the total amount of support from the new high-cost fund will likely not decline materially but will be restructured to better fit a competitive environment.

Low-Income Support for Households. In its May 1997 decision, the FCC concluded that the Lifeline Assistance program, which is designed to mitigate the cost of monthly phone bills, must be offered to qualifying low-income consumers by all eligible telecommunications carriers. In addition, all states are required to provide this service to its consumers.

The new Lifeline programs are scheduled to go into effect on January 1, 1998. For the first time, Lifeline participants will receive an extra $1.75 a month over the $3.50 that Lifeline currently contributes, totaling $5.25 in federal support even if the state provides no funds from the intrastate jurisdiction. In addition, Lifeline will provide matching funds equal to half of the funds generated from the intrastate jurisdiction, up to $7.00 a month in federal support. If a state provides the minimum amount ($3.50 per month) required to elicit the full federal support, the total monthly Lifeline contribution to a reduction in end-user charges could rise from $7.00 under the current system to $10.50 under this new regime.

In addition to services that meet the definition of universal service, Lifeline customers are also allowed to receive toll-limitation service free of charge in areas that this capability is offered. That means that a monthly limit can be set on the amount of money spent on long-distance calling, and if the long-distance bills are not paid, then only the long-distance service, and not the local service, would be cut off until the long distance bill is paid.

In states that provide intrastate support, the eligibility for Lifeline is currently determined by state agencies or telephone companies; this will continue. In states that do not provide intrastate support, however, the default Lifeline eligibility standard will be participation in Medicaid, food stamps, Supplementary Social Security Income, federal public housing assistance, or Low-Income Home Energy Assistance Program. States may choose other criteria, but they must be based solely upon the income of the candidates. These qualifications also apply to Link-Up America, the federal program designed to support the installation of phone lines in homes.

Assistance for Schools and Libraries. Under the 1996 Telecommunications Act, public and nonprofit schools and libraries will be eligible as of January 1, 1998, for discounts applicable to three broad categories of offerings. First, these institutions can procure any telecommunications service on a subsidized basis. Second, support may be received for Internet access. Finally, these discounts can apply to internal networking that is necessary to connect school or library terminals to the Internet. The discounts also cover expenditures for networking hubs, routers, network file servers and server software, and maintenance of network systems. Personal computers, fax machines, modems, and asbestos removal, however, are specifically excluded because they are not considered necessary to transmit information. Training, nonnetwork software, voice mail or information services in general, electrical connections, and security are also not eligible for support.

To receive a discount, applying schools and libraries must seek competitive bids for the eligible services. Price should be the primary factor in accepting a bid, but other factors may be included such as prior experience, personnel qualifications, management capability, and environmental objectives.

The subsidy for authorized services and equipment will be based on whether the school or library is located in a rural or high-cost area as well as the percent of students in the district who are eligible for the national school lunch program that provides free or reduced cost lunches for needy students. Similarly, a library's level of poverty should be calculated on the basis of school lunch eligibility in the public school district in which the library is located. The matrix of discounts developed by the FCC based on the school lunch program and the rural or urban nature of a given school or library ranges from 20% to 90%.

For eligible schools ordering subsidized products or services, the procurement officer for each school district or state applicant is required to certify to the FCC the percentage of students eligible for the national school lunch program. For libraries ordering equipment or services at the library system level, discounts should be computed on an individual branch basis or based on an average of all branches within the system. Where aggregation occurs, the state, district, or system should strive to ensure that each school or library receives the full benefit of the discount to which it is entitled.

Each year, there will be a $2.25 billion cap on universal service support for schools and libraries. Initially, the FCC ruled that funds would be allocated on a first-come, first-served basis, until there is $250 million left in the fund, at which point those schools and libraries that are most economically disadvantaged will be provided funds before others. If the annual cap is not reached in a given year, the unspent funds will be available to support discounts in subsequent years. The Commission subsequently ordered revised rules of priority to ensure that most disadvantaged schools and libraries will receive priority support (Federal–State Joint Board on Universal Service, *Fifth Order* on *Reconsideration and Fourth Report and Order*, cc Docket No. 96-45, rel. June 22, 1998).

All K–12 schools, whether public or private, that operate as a not-for-profit business and have an endowment of $50 million or less are eligible for subsidies. Eligible libraries are as defined in the Library Services and Technology Act, as is "library consortium" with the exception of "international cooperative association of library entities."

To receive subsidies, schools and libraries must:

1. Conduct internal assessments of the components necessary to use effectively the discounted services that they order, including specific plans for using these technologies and integrating them into the curriculum, and receive independent approval of the technology plan, ideally by a state agency that regulates schools and libraries.
2. Submit a complete description of the services that they seek so that it may be posted for competing providers to evaluate.
3. Certify that the school or library is an eligible entity, that the services will be used solely for educational purposes and will not be resold, and that all necessary funding in the current budget year has been allocated to pay for the nonsubsidized portion of the services.

The subsidy is paid directly to the service provider who will then supply the services or equipment to the school or library at the discounted rate.

Assistance for Health Care Providers. Under new FCC rules, health care providers are eligible for universal service support of up to 1.544 Mbps of bandwidth (e.g., T-1 service). This bandwidth may be achieved through any combination of frame relay, ISDN, satellite, or other services whose sum does not exceed 1.544 Mbps. Subsidies will also be provided for any "toll" (long-distance) charges that might be incurred by connecting with an Internet service provider (ISP) who is located outside of the local calling area. Customer premises equipment such as computers and modems are not supported.

Rural health care providers are able to purchase services at rates no higher than that paid by urban providers for similar services. As in the case of schools and libraries, the enabling subsidy is paid to the provider of communication services, who will then provide those services at a discounted rate. The subsidy is calculated on the basis of the "standard urban distance"—that is, the average of the longest diameters of all the cities of a population of 50,000 or more within the state.

Health care providers are required to seek competitive bids for all eligible services by submitting bona fide requests for services to the administrator. The requests shall be posted on the administrator's web site and contain information sufficient to identify the requester and the services requested. The health care provider shall certify that the service chosen is the most cost-effective service available after selection of a telecommunications carrier. The health care provider need not select the lowest bid and may take other considerations into account.

To be eligible for these subsidies, the health care provider must be public or nonprofit and located in a rural area. More specifically, eligible providers include: (a) teaching hospitals or medical schools; (b) community, migrant, or mental health centers; (c) not-for-profit hospitals; (d) local health departments; (e) rural health clinics; and (f) consortia of the previous entities.

There is a $400 million annual cap on the amount of support given to rural health care providers. Collection and distribution of funds began in 1998. Support will be provided on a first-come, first-served basis with the funding requests normally accepted in July before each calendar funding year. In 1997, the timing of the first requests for funding will be a function of the development of an application process for universal service support, which is expected to become available in the fall.

When petitioning for support, all rural health care providers must give a statement to the carrier, signed by an officer of the health care provider who is authorized to order telecommunications services, that certifies the following under oath:

1. The requester is a public or nonprofit entity that falls within one of the seven categories set forth in the definition of health care provider set forth in the Telecommunications Act of 1996.
2. The requested service is supported under these regulations, and the requester is in a rural area or cannot obtain toll-free access to an ISP.
3. The services requested will be used solely for purposes reasonably related to the provision of health care services or instruction that the health care provider is legally authorized to provide under the law of the state in which they are provided.
4. The services will not be sold, resold, or transferred in consideration of money or any other thing of value.
5. If the services are being purchased as part of an aggregated purchase with other entities or individuals, the full details of any such arrangement governing the purchase, including the identities of all copurchasers and the portion of the services being purchased by the health care must be provider.
6. This is the lowest cost method of providing the requested services, taking into consideration the features, quality of transmission, reliability, and other factors that the health care provider deems relevant to choosing an adequate method of providing the required health care services.

Eligible health care providers must file their contracts with the administrator either electronically or by paper. These providers are required to file new funding requests for each new year. Multiyear contracts will be supported on a year-to-year basis, with no lump sum funding permitted. Certifications must be renewed annually.

Eligible health care providers participating in consortia that include (ineligible) private sector members would generally not receive universal service support. The one permitted exception: eligibility would be recognized only if the consortium is receiving tariffed rates, or market rates from those providers who do not file tariffs.

Any public or nonprofit health care provider—regardless of location—who cannot obtain toll-free access to an ISP may obtain either 30 hours of toll-free access or $180 of toll credits (whichever is reached first). Such support can be used to fund toll charges but not distance-sensitive charges for a dedicated connection to an Internet service provider.

Assessment

As in the case of the passage of major telecommunications legislation, the Administration vigorously participated in the CC Docket 96-45 proceeding and provided leadership, in particular, on such issues as the E-rate and rural health care discounts, for both their conceptualization and implementation. For 4 years, the Administration advocated meaningful telecommunications policy reform that would promote competition, spur innovation, reduce prices, and preserve universal service, and "we are well on our way to realizing these goals" with the Commission's adoption of universal service and access charge orders on May 8, 1997 (Secretary of Commerce Daley, 1997). The Vice President remarked that with the new E-rate mechanism, the "nation has taken a great step forward in closing the gap between the information haves and have nots" (Gore, 1997b). The Commission and the Administration agreed on a number of critical elements in the new universal service system, with respect to the rationale that the "most efficient use of the universal service support fund support system should be promoted through the use of market-based techniques wherever possible" (NTIA, 1996b). More specific features strongly supported by both include competitive bidding, tiered discounts based on economic disadvantage, establishing a procurement process that integrates technology with school curricula, and comparability of rural and urban rates for health care providers through toll-free access and distance insensitivity (see Cabinet, 1996, 1997; NTIA, 1996a, 1996b, 1997a, 1997b). It is anticipated that the FCC's actions will have wide-ranging beneficial effects on the future direction and prosperity of this country (Daley, 1997; Gore, 1997b).

SOME CONCLUDING REMARKS

A forward-looking universal service policy should reasonably seek to assure that all who desire access to the NII are able to obtain it. An important part of this policy should be to assure provision of POTS to those households that want but cannot currently subscribe to such service. However, equating universal service solely with telephone service appears to be shortsighted in a society where an individual's economic and social well being depends increasingly on the ability to access, accumulate, and assimilate information through means other than traditional POTS.

A long-term strategy, then, should be to provide a means for electronic—such as Internet—access to all households that desire a connection. In the interim period, schools, libraries, and other community access centers connected to the NII should be part of the public safety net. In terms of connectivity for key institutions identified by the Administration and Congress, good progress has been made. For example, during 1994–1997, the extent of Internet access grew significantly for public schools (123%), public libraries (243%), and hospitals (greater than 127%). By 1997, approximately 78% of public schools, 72% of libraries, and more than 50% of hospitals could claim access to the information superhighway. (See NCES, 1997; NCLIS, 1997b; RUS, 1996).

Partnerships—intergovernmental as well as those involving both the public and private sectors—should play a pivotal role in designing and implementing a new universal service policy. For example, the aforementioned Federal–State-Local Summit permitted the views of the three levels of government to be discussed and—where possible—harmonized, and these principles were largely captured in the Telecommunications Act of 1996. The Administration's five field hearings and virtual conference afforded a broad spectrum of interests to provide input at this critical policymaking juncture.

As a complement to universal service, the TIIAP grant program has proven to be effective in extending the benefits of advanced telecommunications technologies to millions of Americans in rural and underserved areas through matching grants and public–private collaborations. Similarly, grants awarded by the Department of Education and Rural Utilities Service have helped schools, libraries, and rural health care providers to meet the demands of the burgeoning Information Age. As directed by the 1996 Telecommunications Act, the FCC has collaborated with a specially created Federal–State Joint Board to develop detailed policies regarding subscribership, defining universal service, and funding. Effective partnering with respect to these and other types of initiatives that address training, technical support, maintenance, and user-friendly applications as well as connectivity will be crucial to the nation's future prosperity.

Good public policy requires a solid factual predicate. As discussed in this chapter, recent research has revealed new insights about information have nots. For example:

1. The overwhelming majority of Americans want (and currently have access to) telephone service. Although some persons do not desire basic phone service, most who are phoneless are apparently unable to subscribe for one or more socio-economic reasons.

2. Those households who do not have phones and/or online access are disproportionately low income, minority, young (also, old in the case of PCs/modems), less-educated, mobile, and located in central cities or rural areas. Importantly, many of the disadvantaged are the most enthusiastic users of online services that facilitate economic uplift and personal empowerment.
3. For many, staying connected to the network may be a bigger problem than achieving an initial hookup, either for POTS or the Internet.

However, more work needs to be done regarding the have nots. For example, are the same low-income, disadvantaged individuals also those who are minorities, less-educated, young, or mobile? What is the relationship, if any, between low income and high-cost areas for schools and libraries vis-à-vis households? Is flat-rate long-distance connectivity to the Internet good policy in rural areas? Will Internet telephony—with lower user prices at least partly attributable to an exemption from payment of subsidy-based access charges—help or harm overall subscribership? As a complement to the new universal service policy, can improved access to training and technical support reduce the incidence of Internet dropouts?

Invoking the eternal lament of those who do research, this chapter must conclude by exhorting analysts to bring into sharper focus the true profile of the information disadvantaged. Only when this is achieved can policymakers put all Americans on the road to the 21st century.

REFERENCES

AT&T (1982). *Bell System statistical manual.* New York: AT&T Comptroller.

Belinfante, A. (1998). *Telephone subscribership in the United States.* Washington, DC: Federal Communications Commission.

Bolter, W., McConnaughey, J., & Kelsey, F. (1990). *Telecommunications policy for the 1990s and beyond.* Armonk, NY: M.E. Sharpe.

Brown, K. (April 12, 1996). *Testimony Before the Federal–State Joint Board on Universal Service* (CC Docket No. 96-45).

C&P Telephone Company. (1993). *Telephone penetration project: Disconnect study.* Washington, DC: C&P of the District of Columbia.

Cabinet. (1996, October 10). Letter to FCC Chairman R. Hundt from Secretaries R. Riley (U.S. Department of Education), D. Glickman (U.S. Department of Agriculture), and M. Kantor (U.S. Department of Commerce), In the Matter of Federal–State Joint Board on Universal Service (CC Docket No. 96-45).

Cabinet. (1997, April 28). Letter to FCC Chairman R. Hundt from Secretaries D. Glickman (U.S. Department of Agriculture), W. Daley (U.S. Department of Commerce), and D. Shalala (U.S. Department of Health and Human Services), In the Matter of Federal–State Joint Board on Universal Service (CC Docket No. 96-45).

Clinton, W. (1994, January 25). *President's State of the Union Address*, Washington, DC.

Clinton, W. (1996a, January 23). *President's State of the Union Address*, Washington, DC.

Clinton, W. (1996b, February 1). *Statement by the President*, Washington, DC.

Clinton, W. (1996c, February 8). *Remarks by the President in signing ceremony for the Telecommunications Act Conference Report*, Washington, DC.

Clinton, W. (1997, January 20). *President's Inaugural Address*, Washington, DC.

Clinton, W. (1998a, January). *President's State of the Union Address*, Washington, DC.

Clinton, W. (1998b, June 5). *Remarks by the President at Massachutsetts Institute of Technology 1998 Commencement*, Cambridge, MA.

Federal Communications Commission. (1997). *Trends in telephone service*. Washington, DC: Author.

Field Research Corporation (1993). *Affordability of telephone service: Non-customer survey*. los Angeles: GTE-PacTel.

Gonzalez, E. (1995). *Connecting the nation: Classrooms, libraries, and health care organizations in the information age*. Washington, DC: U.S. Department of Commerce.

Gore, A. (1994, January 11). *Vice President's Remarks before the Academy of Television Arts and Sciences*, Los Angeles, CA.

Gore, A. (1995, January 9). *Vice President's Remarks before the Federal–State-Local Telecommunications Summit*, Washington, DC.

Gore, A. (1996a, February 2). *Statement of the Vice President on Passage of Telecommunications Reform Legislation*, Washington, DC.

Gore, A. (1996b, June 29). *Vice President's Remarks, NetDay96*, Washington, DC.

Gore, A. (1997a, February 8). *Joint Radio Address with President Clinton.*

Gore, A. (1997b, May 7. *Statement of the Vice President.*

Gore, A. (1998a, April 29). Statement.

Gore, A. (1998b, May 8). *Essay*. Markle Foundation's e-mail for all online conference.

Horrigan, J., & Rhodes, L. (1995). *The evolution of universal service in Texas*. Austin, TX: Lyndon Baines Johnson School of Government, University of Texas.

Information Infrastructure Task Force (IITF) (1993, September 15). *National information infrastructure: Agenda for action*. Washington, DC: IITF.

International Telecommunication Union (ITU). (1997). *World telecommunication development report: Universal access 1998*. Geneva, Switzerland: ITU.

Irving, L. (1994, November 14). *Keynote Address*, NTIA Virtual Conference.

Irving, L. (1997a, February 7). *Remarks, NetDay 2000 Conference*, Washington, DC.

Irving, L. (1997b, April 24). *Testimony on Reauthorization of NTIA Before the Subcommittee on Telecommunications, Trade, and Consumer Protection, Committee on Commerce, U.S. House of Representatives*. Washington, DC: NTIA.

Irving, L. (1998, May 8). *Essay*, Markle Foundation's e-mail for all online conference, held May 4–15, 1998.

Katz, J., & Aspden, P. (1997). *Internet dropouts: The invisible group*. (Online). Markle Foundation Home Page http://www.markle.org>: Bellcore.

Lande, J. (1993). *Reference book: Rates, Price Indexes, and Household Expenditures for Telephone Service*. Washington, DC: Federal Communications Commission.

McConnaughey, J., Nila, C., & Sloan, T. (1995). *Falling through the net: A survey of the "have nots" in rural and urban america*. Washington, DC: National Telecommunications & Information Administration.

Mueller, M., & Schement, J. (1996). Universal service from the bottom up: A profile of telecommunications access in Camden, New Jersey, *Information Society, 12* (273–292).

National Center for Education Statistics (1996, February). *Advanced Telecommunications in U.S. Public Elementary and Secondary Schools, 1995*, NCES 96-854.

National Center for Education Statistics (NCES) (1997, February). *Advanced Telecommunications in U.S. Public Elementary and Secondary Schools, Fall 1996*, NCES 97-944.

National Center for Education Statistics (NCES). (1998, February). *Internet Access in Public Schools (Issue Brief).*

National Commission on Libraries and Information Science (NCLIS). (1994). *Public libraries and the Internet: Study results, policy issues, and recommendations.* Washington, DC: NCLIS.
National Commission on Libraries and Information Science (NCLIS). (1997a). *The 1996 national survey of public libraries and the Internet.* Washington, DC: NCLIS.
National Commission on Libraries and Information Science (NCLIS). (1997b). *The 1997 national survey of public libraries and the Internet.* Washington, DC: NCLIS.
National Education Association and Princeton Survey Research Associates. (1993). *Technology in the classroom: A teacher perspective.*
National Telecommunications and Information Administration (NTIA)/U.S. Department of Commerce. (1994a). *Inquiry on universal service and open access, 59 Fed. Reg. 48112, 4814-15 (NTIA NOI).* Washington, DC: Federal Register.
National Telecommunications and Information Administration (NTIA)/U.S. Department of Commerce. (1994b). *NII field hearings on universal service and open access: America speaks out.* Springfield, VA: National Technical Information Service.
National Telecommunications and Information Administration (NTIA)/U.S. Department of Commerce. (1994c). Virtual conference archives. (Online). NTIA Web Site. Available at http://www.ntia.doc.gov.
National Telecommunications and Information Administration (NTIA)/U.S. Department of Commerce. (1995). *Survey of rural information infrastructure technologies.* Boulder, CO: Institute for Telecommunication Sciences.
National Telecommunications and Information Administration (NTIA)/U.S. Department of Commerce. (1996a). *Reply Comments of the National Telecommunications and Information Administration, In the Matter of Amendment of the Commission's Rules and Regulations to Increase Subscribership and Usage of the Public Switched Network* (CC Docket No. 95-115).
National Telecommunications and Information Administration (NTIA)/U.S. Department of Commerce. (1996b). *Reply Comments of the National Telecommunications and Information Administration, In the Matter of Federal–State Joint Board on Universal Service* (CC Docket No. 96-45).
National Telecommunications and Information Administration (NTIA)/U.S. Department of Commerce. (1997a). *Reply Comments of the National Telecommunications and Information Administration, In the Matter of Federal–State Joint Board on Universal Service* (CC Docket No. 96-45).
National Telecommunications and Information Administration (NTIA)/U.S. Department of Commerce. (1997b).
(Online) NetDay 2000 Home Page. Available at: http://www.netday.org
Office of Technology Assessment. (1995). *Teachers and technology: Making the connection,* OTA-EHR-616. Washington, DC: Government Printing Office.
Quality Education Data, Inc. (1995). *Technology in Public Schools, 1994–95,* Denver, CO: QED, Inc.
Rural Utilities Service (RUS). (February 1996). *Bringing the information superhighway to rural america* [Online]. RUS Web Site. Available at: http://www.usda.gov/rus/home
Secretary of Commerce Daley, W., (May 9, 1997). *Statement.*
Brown, R.,(1995, January 9). *Remarks at the Federal–State–Local Telecom Summit.*
Schement, J., Belinfante, A., & Povich, L. (1994). *Telephone penetration 1984–1994.* Washington, DC: Project on Information Policy, Rutgers University.
TIIAP, Fact Sheet (May 1998).
Southwestern Bell Telephone. (1986). *Kansas disconnect study.* Kansas City: Author.
Tech Corps, "A One Page Perspective." [Online] Web site: http://www.ustc.org
21st Century Teachers Network (1997). (Online) Available at: http://www.21ct.org
U.S. Bureau of the Census/U.S. Department of Commerce. (1994a). *Computer ownership/usage supplement* and *current population survey.* Washington, DC: U.S. Bureau of the Census.

U.S. Bureau of the Census/U.S. Department of Commerce. (1994b).*Statistical brief: Phoneless in America*, CB94-127, and database. Washington, DC: U.S. Bureau of the Census.

U.S. Bureau of the Census/ U.S. Department of Commerce. (1997). *School enrollment supplement and current population survey*. Washington, DC: U.S. Bureau of the Census.

U.S. Department of Agriculture (USDA). (1997). *News Release No. 0291.97*. [Online]. USDA Home Page. Available at: http://www.usda.gov

U.S. Department of Education. (1996). *Getting America's students ready for the 21st Century: Meeting the technology literacy challenge*.

Weinhaus, C., Makeef, S., Jamison, M., et al. (1992). *Who pays whom? Cash flow for some support mechanisms and potential modeling of alternative telecommunications policies*. Cambridge, MA: Harvard University Program on Information Resources Policy.

White House. (1996, April 17). *Executive Order on Educational Technology*.

White House. (1997). *The President's educational technology initiative*. [Online]. White House Web Site. Available at: http://www1.whitehouse.gov/WH/EOP/OP/html/edtech

Wilhelm, T. (1996). *Latinos and information technology: Perspectives for the 21st century*. Claremont, CA: The Tomas Rivera Center.

VI

THE ROLE OF THE STATES

14

The New State Role in Ensuring Universal Telecommunications Services

Thomas W. Bonnett
Public Policy Consultant, Brooklyn, NY

The federal Telecommunications Act, which President Clinton signed into law on February 8, 1996, shattered the social compact that had governed the provision of basic telephone services in this country for most of this century. Beginning in 1907, state regulation of local telephone services had evolved into an elaborate social compact that granted special privileges and responsibilities to the local exchange companies (Cohen, 1992). The antecedents of the social compact were public service companies established in 19th-century U.S. law, such as railroads and grain elevators, as well as English common law (Stone, 1989).

The explicit terms of the modern social compact in the telephone industry included: the granting of an exclusive monopoly franchise to the local exchange company; the public regulation of both the prices and quality of services to protect consumers; and a rate-setting process that granted reasonable profits to the company's shareholders to ensure that adequate infrastructure investment was maintained (Stone, 1989). The highly evolved social compact also included the responsibility of the monopoly provider to provide various public service obligations. As defined by Noam (1994a), the *universal service obligation* requires a carrier to reach every willing user and desired destination wherever located whereas *common carriage* is the obligation to provide services without discrimination to all users, given a physical plant. A related concept is *carrier of last resort (COLR)*, which is the requirement to provide services to all customers within a service area (Borrows, Bernt, & Lawton, 1994). Hence, granting an exclusive monopoly franchise to the local exchange company along with public regulation of its prices, profits, and infrastructure investments constituted a social compact that was generally accepted by the public for most of this century (Teske, 1995).

The policy regime of publicly regulated monopoly providers of essential services, such as telephone services and electricity, had distinct advantages. The stability and security of this regime appealed to ratepayers, stockholders, and employees. Regulators enjoyed the opportunity to set rates for some services at levels higher than actual cost (e.g., business rates, intrastate toll calls, interstate toll calls), which generated sufficient revenue to keep residential rates affordable (Crandall, 1991). This latter objective—low residential charges for basic telephone services—has been popular with the general public and most regulators (Teske, 1995). It also became associated with the concept of universal service, usually defined as the social goal of ensuring that basic telephone services are made available to everyone in society (Noam, 1994b). The transition from the policy regime of publicly regulated monopoly providers into the bold, new regime of competition poses significant challenges to federal regulators and state policymakers. The policy challenge is to create a level playing field for all competitors in the liberalized markets without forsaking the social objectives of ensuring universal service.

This chapter addresses the policy challenge faced by the states to ensure universal service amidst emerging competition by new entrants in liberalized markets. It outlines how competition driven by cost will erode existing rate structures, defines the rationale for public efforts to achieve universal service, and discusses the state role in preserving universal service in the context of the federal Telecommunications Act of 1996. The emerging state role on universal service is the most significant challenge that the states have faced in telecommunications policy during the last century.

THE TELECOMMUNICATIONS ACT OF 1996 AND UNIVERSAL SERVICE

When Congress enacted the Act, it endorsed a bold, new paradigm for public policy for telecommunications. The federal legislation sought to promote competition in telecommunications markets, deregulate much of the telecommunications industry, and ensure universal service. To advance the brave, new world of competition, the legislation imposed responsibilities on incumbent local exchange companies, retained roles for the Federal Communications Commission (FCC) and state commissions to ensure that a level playing field was established, and charged the FCC with eliminating the implicit subsidies hidden in the regulated rate structures for interstate long-distance services. To achieve universal service objectives, the legislation created a new package of direct subsidies to all qualified providers.

The following themes are clearly reflected in the aftermath of this bold legislation. They are manifested in the specific issues that will be resolved, perhaps through the courts, in subsequent years: the jurisdictional conflict between the FCC and the states; the transition from the explicit social compact (of the old regime) toward one in which competitive neutrality yields equal rights and burdens to all carriers; and the policy formulation of articulating new social goals—and the mechanisms for achieving them—in a competitive regime.

GENERAL RATE ISSUES FACING POLICYMAKERS

In a competitive market, prices would be set, theoretically, by competitive producers according to their marginal costs[1]; this is a paradigm shift from the regulated environment that employed rate distortions to keep basic rates affordable. Crandall (1991) presented a summary of the distortions in traditional telephone rate policy:

> Long-distance rates were held artificially high so as to mitigate increases in local rates. Long-distance rates were based on distance, not on call density. Local rates were usually lower in high-cost rural areas than for lower-cost urban areas. Business users were charged more for access and local exchange service than residential users in the same exchange. Local service was generally offered on a flat-rate basis; therefore, heavy users paid no more for local service than users who placed less of a burden on traffic-sensitive facilities. Each of these distortions invited some form of competitive entry if the regulators could be persuaded to allow it. (p. 23)

The old regulated rate structure will eventually collapse under the onslaught of competition that drives prices toward actual costs. Will prices change much? Which prices will change greatly? Which ones only slightly? Will these pricing changes affect universal service objectives? How much will household penetration rates be affected if prices replace tariffed rates? What direct subsidies might be needed to ensure universal service in a regime in which prices are set by market forces and not by rate regulators?

An issue analysis published by the Citizens for a Sound Economy, a conservative nonprofit organization, estimated that roughly $17.5 billion is diverted by the rate distortions identified by Crandall and spent on federal direct subsidy programs (Leighton, 1995). The direct subsidy programs include the Universal Service Fund (USF), the National Exchange Carriers Association (NECA) funds, and low interest loans from the Rural Electrification Administration and Rural Telephone Bank. The direct subsidy from the USF is approximately $776 million, which is assessed against the interexchange (long-distance) companies. According to this analysis, the distortions within rate structures represent a much greater redistribution of resources than all of the direct subsidy programs.

State regulators will not have an easy time navigating between the shoals of emerging competition and the rocks of universal service goals.

---

[1]Telecommunications is a declining cost industry because the fixed investments of building the networks are so high. Services priced at marginal costs would not be sustainable because the revenues would not provide for the recovery of the facility costs of building the network infrastructure. Consequently, competitive pricing for various services will, over time, approach average rather than marginal cost.

The National Regulatory Research Institute, the research arm of NARUC, uses a similar metaphor: "Regulators are like the captain of a ship with a universal service engine propelling the ship in one direction, while a competition engine pushes the ship in another direction" (Conte, 1995, p. 32). Mueller (1995) added a sharper observation: "There is something a bit ridiculous about a policy regime that tries to undermine subsidies with one hand (by promoting competition) and save them with the other" (p. 227). A report of the Governor's Telecommunications Policy Coordination Task Force in Washington State (1996) includes a good summary of the policy challenge ahead.

> A variety of subsidies have evolved with the telecommunications industry. Enhanced 911, the Telephone Relay Program for the hearing impaired, the Telephone Assistance Program for low-income residents, and universal service pools are explicit subsidies. *Others are implicit, such as the use of statewide average rates and toll access rates to even out the cost of service among customers. While some of these subsidies have helped provide essential public services and should continue in force, other subsidies have been called into question by increasingly heated competition. It may no longer make sense for some ratepayers to defray the costs of providing services to other ratepayer groups.* The WUTC needs to review these subsidies and determine whether these subsidies are being applied in a competitively neutral manner. The Washington State Utility and Transportation Commission (WUTC) should report back to the Governor, Task Force and Legislature with recommendations for rationalizing or eliminating these subsidies. (p. 5, italics added)

Many analysts anticipate that rates will be fundamentally restructured as a result of competition in local and intrastate services. This is called rate deaveraging or rate rebalancing. Two Regional Bell Operating Companies (RBOCs) sought to rebalance rates in 1996, increasing rates in rural areas to reflect the higher cost of providing service: US West in Washington State and Bell Atlantic in Pennsylvania. Cincinnati Bell has proposed a rate rebalancing "that would shift $70 million in revenues from basic local business services onto residential service" (Kirchhoff, 1997a, p. 8). Other incumbent providers will seek to rebalance rates in anticipation of, or in response to, emerging competition in local telephony.

However, state regulators have resisted rate rebalancing, arguing that it is premature or not justified. For example, in the spring of 1996, the Washington State Utility and Transportation Commission (WUTC) rejected the rate proposal by US West to increase rates for its rural customers, ruling that residential rates are priced above cost on an average statewide basis (Governor's Telecommunications Policy Coordination Task Force, 1996).

According to Gabel (1995a), the state commissions in Colorado, New Hampshire, Florida, Pennsylvania, and Kentucky have also ruled in recent years that basic residential service is not subsidized. A full discussion of this interpretation is beyond the scope of this chapter, but the significant problem with this conclusion is that it is made based on an aggregate, not a deaveraged, analysis. For example, the cost of providing basic service through most rural wire centers is always much higher than the averaged tariffed rate, whereas the cost at most urban wire centers is much lower than the tariffed averaged rate for the incumbent provider.[2] Whether basic residential services—on a statewide basis—pay for themselves, as the consumer advocates suggest, or whether they benefit from various cross subsidies, as most economists contend, has much less significance in the future if robust competition in lucrative markets drives all prices toward actual costs. That outcome—driven by market forces—will eviscerate the long-standing practice of rate averaging, forcing rate rebalancing.

However, the old regime of publicly regulated monopoly providers—employing rate averaging—was the conventional approach to provide what was generally regarded as essential services to the public. As that regime crumbles, the immediate public policy challenges are: how to define the services that should be supported; how to design the system of direct subsidies to support those services; and how to achieve competitive neutrality among all carriers in the network system.

DEFINING UNIVERSAL SERVICE AND AFFORDABILITY UNDER THE 1996 ACT

The Act directed a joint board—consisting of members of the FCC and state regulators, and a consumer advocate—and the FCC to assume the unenviable task of trying to define what is meant by universal service. Section 254 makes it clear that the term is not a static concept: "Universal service is an evolving level of telecommunications services that the Commission shall establish periodically under this section, taking into account advances in telecommunications and information technologies and services." Before our society entered the digital age, access to plain old telephone service (POTS) was the conventional standard. Today an array of digital signals—including voice, pictures, data, and text—can be transmitted through various telecommunications networks to the average consumer. This provision of the law ensures that the definition of universal service will evolve "in the future world of integrated digital networks and multimedia ... beyond basic analog telephone service" (Egan, 1996, p. 225). The future role for states will center on whether telecommunications services beyond basic telephone services should be included in deliberations on universal service definitions and programs.

The decision rule for the new definition of universal service is stated in Sec. 254 (c)(1):

---

[2]The author thanks Barbara Cherry for this insight.

> The Joint Board in recommending, and the Commission in establishing, the definition of the services that are supported by Federal universal service support mechanisms shall consider the extent to which such telecommunications services—
>
> (A) are essential to education, public health, or public safety;
> (B) have, through the operation of market choices by customers, been subscribed to by a substantial majority of residential customers;
> (C) are being deployed in public telecommunications networks by telecommunications carriers; and
> (D) are consistent with the public interest, convenience, and necessity.

The Act also included the following Universal Service Principles in Sec. 254(b) to guide the Joint Board and the Commission: that quality services should be available at just, reasonable, and affordable rates; that access to advanced telecommunications and information services should be provided to all regions of the nation; that consumers in all regions of the nation, including low-income consumers and those in rural, insular, and high-cost areas, should have access to telecommunications and information services and rates that are reasonably comparable to those services provided in urban areas; that all providers of telecommunications should make an equitable and nondiscriminatory contribution to the preservation and advancement of universal service; that there should be specific, predictable, and sufficient federal and state mechanisms to preserve and advance universal service; that certain educational institutions, health care providers, and libraries should have access to advanced telecommunications services; and such other principles as the Joint Board and the Commission determine are necessary and appropriate for the protection of the public interest, convenience, and necessity and are consistent with the Act.

The Joint Board issued its recommendations on November 8, 1996, and the FCC issued its order in May 1997. In its order, the FCC adopted both an additional principle of universal service, competitive neutrality, and a definition of universal service, as recommended by the Joint Board. Universal service is defined to include the provision of the following core services to primary residential lines and single-line businesses: voice grade access to the public switched network; dual-tone multifrequency signaling (i.e., touch tone); single-party service; access to emergency services; access to operator services; access to long-distance services; and access to directory services. Low-income customers are also to have access to toll blocking and toll limitation service free of charge.

In its order, the FCC also directed the states to monitor rates and nonrate factors, such as subscribership levels, to ensure affordability. Yet, the FCC deferred establishment of a new federal high cost fund until 1999. State policymakers are acutely concerned about whether the ultimate federal universal service programs will be sufficient to ensure that basic services remain affordable to everyone regardless of geography. If the federal high cost fund is too low, elected state officials from predominantly rural states fear that rural rates will climb substantially or that state universal service funds, established to achieve the same redistribution function, will be large, expensive, and burdensome to the providers and the general public.

In determining affordable rates in Section 254(b)(1), Cooper (1996) recommended that state regulators define "affordable" in a relative sense: "If it hurts a lot to pay for telephone service, telephone service is not deemed to be affordable, even though the subscriber continues to pay for it" (p. 10). On the other hand, some experts encourage state regulators to resist rate rebalancing by citing Section 254(b)(3), which refers to the aspiration that low-income consumers and residents in high-cost areas should have access to services that are "reasonably comparable to those services provided in urban areas and that are available at rates that are reasonably comparable to rates charged for similar services in urban areas" (Gabel, 1996a).

The two interpretations of affordability under the Act pose an almost impossible mission for regulators. In the old regime of regulated monopoly providers, rate distortions kept residential rates affordable. In the new regime, prices will be set by market forces, not through regulated tariffs. Similarly, state regulators used rate averaging to produce reasonably comparable services at reasonably comparable rates across areas with divergent cost structures. This new law encourages competitive markets to set prices and then compensates providers with a program of direct subsidies to ensure reasonably comparable services and rates for rural areas. Can services and rates be kept—through direct subsidies—at reasonably comparable levels across geographies with extreme cost variations among them?

A few states have begun to address this question by providing a range within which rates are presumed to be valid. For example, Bernt (1996) noted "California policy is to require that rural rates are no higher than 150 % of urban rates, and rates for low income customers are no less than 50 % of urban rates" (p. 73). With the emergence of competition in local telephony developing in various markets at an uneven pace and the prospect of rate rebalancing occurring as a result, this is likely to be a difficult challenge for many state regulators.

Adopting a rate range standard, however, presents a paradox because the current range in rates for basic services is already substantial. Although most consumers are served by a Bell Operating Company, as many as 20% of consumers are served by independent telephone companies. Just four states are served by one provider: Delaware, District of Columbia, Hawaii, and Rhode Island. Hence, telephone subscribers in all other states are subjected to different rates depending on which incumbent provider serves them, the service territory of the provider, and the underlying cost structure for service delivery. Based on a study by Goldstein and Gooding (1995), the range in residential rates for the independent telephone companies was greater than the range for the Bell Operating Companies. Within states, rates for independent telephone companies included the monthly ranges shown in Table 14.1.

The data in Table 14.1 should be interpreted carefully. First, the basic services offered and the quality of services are likely to vary among providers. Second, in most cases, relatively few subscribers are served by these independent companies. Nevertheless, the data show that most states had rather broad ranges of rates for residential services prior to the enactment of the Act and its new focus on ensuring affordability and reasonably comparable services and rates. With this as an empirical background, state regulators will not have an easy time assessing reasonably comparable services at reasonably comparable rates.

Given the tension between the pressure to rate rebalance and the directive to keep rates affordable, the transition toward competition in local telephony creates a compelling argument for establishing state universal service funds to complement federal support programs. As a result, key public policy questions will challenge state elected officials and regulators in the near term, as the FCC implements its new universal service order. For example, will the federal universal service programs ensure that everyone, regardless of location, has access to basic services at affordable prices? Will the states embrace competition through advantageous prices for interconnection and wholesale rates? Will competition in local telephony stimulate private investment, improve quality, and lower prices; alternatively, will competition erode the revenue base that is necessary to fund infrastructure improvements, preserve the network, and retain quality services? What will be the effect of competition in urban and in rural areas? Is rate rebalancing inevitable? Will states choose to establish universal service funds to supplement the federal universal service programs? Can the new competitive regime achieve competitive neutrality by establishing symmetrical regulations relating to universal service support, funding, and requirements?

DIFFICULTY IN ACHIEVING COMPETITIVE NEUTRALITY

The tradition and legacy of regulating monopoly providers pose a new set of policy challenges for state regulators in the new regime of promoting competition in local telephony. State officials must begin to define policy in terms of what is best for the industry and the consumers, rather than negotiating with incumbent providers on behalf of the public. Similarly, states must follow the principle of competitive neutrality, which the FCC adopted in its order. Any action that is perceived to restrict competitive entry is subject to federal preemption by the FCC under Sec. 253.

TABLE 14.1

Monthly Range of Rates for Independent Telephone Companies

State	*Low rate*	*High rate*
Arizona	$4.50	$30.00
Arkansas	$5.00	$32.00
Colorado	$4.15	$30.00
Georgia	$4.00	$18.00
Illinois	$5.40	$28.00
Indiana	$3.00	$25.00
Iowa	$2.00	$24.78
Kansas	$3.50	$13.00
Kentucky	$5.00	$18.00
Louisiana	$9.00	$18.50
Maine	$4.75	$14.50
Michigan	$3.76	$12.30
Minnesota	$5.00	$30.00
Missouri	$4.00	$16.00
Nebraska	$4.00	$15.00
Nevada	$5.75	$16.00
New York	$3.00	$17.92
North Carolina	$2.56	$18.26
Ohio	$2.70	$22.90
Oklahoma	$5.00	$20.00
Oregon	$8.00	$16.00
Pennsylvania	$3.25	$17.73
South Carolina	$3.00	$16.90
South Dakota	$5.25	$15.75
Texas	$5.05	$19.00
Virginia	$6.00	$16.00
Washington	$7.00	$26.00
Wisconsin	$2.90	$25.00

Source: Goldstein and Gooding, 1995.

The Act and most state statutes still retain a host of special responsibilities, imposed traditionally on incumbent local exchange carriers, from the old regime. Cherry and Wildman (chap. 3, this volume) demonstrate the difficulty in achieving competitive neutrality by taking a close look at the various responsibilities still imposed on providers. Some of these obligations include common carrier obligations, price regulation, earnings regulation, and the carrier of last resort obligations. The authors conclude that asymmetric unilateral rules (e.g., obligations imposed only on incumbent providers but not new entrants) are not sustainable in the new competitive regime. In short, the new competitive regime requires that all public service obligations either be imposed equally on all providers, or, if imposed unilaterally on one provider, must be accompanied by adequate compensation as part of a bilateral rule.

The challenge of achieving competitive neutrality in the design of universal service programs will rest, in part, with state officials. For example, the state commission has the responsibility to designate intrastate eligible telecommunications carriers (ETC) under Sec. 214 of the Act; these are common carriers of services eligible to receive funds under universal support mechanisms. To date, only incumbent telephone companies serving high cost areas have been eligible to receive assistance through universal service funds. In the past, monopoly providers were granted the exclusive franchise to serve a designated area; in return for this exclusive franchise, the local exchange carrier (LEC) would assume public service obligations, including serving users without discrimination (common carriage) and serving all customers within the service area (COLR; Bernt, 1996).

In the competitive regime, any carrier could be designated as an ETC by the state commission, and thereby eligible to receive universal support funding, provided it meets certain statutory criteria. For example, the Act "requires state commissions to designate more than one ETC for every service area, except those areas served by small rural telephone companies" (Bernt, 1996). If found to be in the public interest, the state commission may designate more than one ETC in an area served by a rural telephone company; in all other service areas, the state commission must designate more than one ETC.

State decisions regarding the service areas and other requirements of the ETCs will affect the complexity of the universal service program and determine the range of choices of providers by consumers. The amount of direct subsidy needed is the difference between the cost of providing basic services and the lesser amount of revenues earned by the provider. Once specified, the direct subsidy would flow to any ETC. A direct subsidy that is too generous provides a competitive advantage to the receiving ETC; a subsidy that is too low would discourage any carrier from becoming an ETC and providing universal services (Bernt, 1996).

The states will vary in how they define service areas for the ETCs. According to Bernt (1996), the Ohio PUC "allows new entrants to self-define their service areas, and to receive universal service subsidy funds for those areas" (p. 49). In contrast, several states legislatures in 1997 considered proposed legislation that would require ETCs to accept the same service areas as being currently served by the incumbent local exchange carriers. Bernt (1996) explained the policy implications concerning how service areas for the ETCs are defined by the states:

> Allowing ETCs to serve only parts of the ILEC's territory will place the ILEC at a disadvantage and allow the competing ETCs to "cream skim," to serve those segments of the ILEC territory with the densest population and the easiest terrain. The ILEC meanwhile would be forced to serve all of the ILEC's service territory. On the other hand, having to serve all of the ILEC's service territory could present a significant barrier to entry for competing ETCs, few, if any, of whom would have facilities and resources in place in that whole territory. (p. 49)

STATE UNIVERSAL SERVICE FUNDS

California established a Lifeline Fund in 1983 to subsidize basic telephone services to low-income households (Rosenberg & Stanford, 1996). Rhode Island followed a year later with its Lifeline Fund.

Then in 1984 California established a High Cost Fund, which provided direct subsidies to companies serving rural and high cost areas. Because the cost of providing basic services was much higher than the statewide average, these direct subsidies to the companies served to reduce basic rates below actual costs. Illinois established its high-cost fund in 1986. Only one or two states established similar funds each year until the mid-1990s when it appeared that Congress might enact major legislation on telecommunications.

During the 1994 and 1995 legislative sessions, 12 states enacted provisions to ensure universal service access, usually by requiring all telecommunication providers to contribute to a universal service fund.[3] The design and scope of these state programs vary, although most serve to expand access to basic services to low-income households (usually matching the federal Lifeline program) or to customers in rural and high-cost areas (through direct subsidies from high cost funds).

The Act explicitly authorizes states to create their own universal service programs. Section 254(f) states: "A State may adopt regulations not inconsistent with the Commission's rules to preserve and advance universal ser-

---

[3]The following states enacted legislation in 1994 and 1995 relating to universal service: Colorado, Connecticut, Florida, Georgia, Hawaii, Iowa, Minnesota, North Carolina, Tennessee, Texas, Wisconsin, and Wyoming (Itkin & McLaughlin-Krile, 1995).

vice." All telecommunications carriers that provide intrastate services shall contribute, on an equitable and nondiscriminatory basis, in a manner determined by the state to preserve and advance universal service. The statute also enables states to adopt additional definitions and standards for universal service as long as the state regulations specify specific, predictable, and sufficient mechanisms and do not rely on or burden federal universal service support mechanisms (Bonnett, 1996).

Among the telecommunications issues confronting state policymakers, this one is at the top of the chart. The states have the authority to adopt universal support programs that are consistent with the FCC's rules, including establishment of a higher standard of universal service than the federal one. State universal service funds could provide assistance to supplement federal support mechanisms in order to meet such higher service needs.

The constraint on the states is largely a political one: How much would state policymakers want to charge the providers of local and intrastate telephone services or each subscriber to fund various programs to achieve universal service objectives? The fundamental conflict over the redistribution of social resources to achieve universal service objectives, which is a political debate, is inherent in formulating state policies toward universal service programs and funding.

Rosenberg and Stanford (1996) provided a summary of state actions with regard to universal service policy. For example, most states define universal service as basic service, but their definitions vary considerably. How these definitions vary will be discussed later.

More than 10 states provide direct subsidies to customers in high-cost areas; a few states limit support to qualified low-income customers (Lifeline); and some states provide support in some cases if both customers and service areas meet established criteria.[4]

As to funding mechanisms, most states levy a surcharge on telecommunications revenues, usually minus payments to other providers, to fund their universal service programs. Six states have authorized a common line charge, which is paid by each subscriber, to fund these programs. Nine states use other charges (such as a charge on intrastate toll minutes of use). Furthermore, most states collect these revenues from all telecommunica-

[4]As of May 1996, according to Rosenberg and Stanford (1996), the following states provide universal service support in defined service areas: Arizona, California, Colorado, Idaho, Illinois, Kansas, Maine, Oklahoma, Pennsylvania, and Vermont. The survey identified four states that provided universal service support based on customer eligibility: California, Illinois, Rhode Island, and Texas. Four states determined eligibility for universal service support by service areas and customers: Hawaii, Oregon, Wisconsin, and Washington state. Five states employed other criteria for their universal service support mechanisms: Connecticut, Georgia, Indiana, Tennessee, and Wyoming.

tions providers, although three states limit liability to carriers providing toll calls.[5]

Estimating the cost of universal service and appropriate funding for high-cost support to providers is a difficult and controversial task. Using the actual cost data of incumbent providers as the basis for the cost of providing universal service in high-cost areas is problematic because it includes embedded costs; using proxy models to estimate the cost of providing universal service is suspect because any model requires a host of assumptions about the cost of inputs, the level of services to be provided, and various technologies employed to provide basic services (Gabel, 1996b). Holding auctions to determine the price at which providers are willing to serve as COLR has, in theory, the disadvantage of encouraging providers to bid at a such a low price that they are incapable of delivering quality services (Bernt, 1996). That concern has not prevented the FCC or the California Public Utility Commission from considering the future use of an auction mechanism. According to Vogelsang and Mitchell (1997), auctions have been used in Australia to select COLRs.

TARGETED VERSUS EXPANDED STRATEGIES TO UNIVERSAL SERVICE

Prior to the 1960s, POTS at affordable rates was one of the standards for universal service. In recent years, many states have defined basic services to include: affordable rates, touch-tone dialing, access to long-distance carriers, and 911 emergency services. Vermont's broad-based universal service fund, which is financed by a 1.25% surcharge on each customer's bill collected by all service providers, generates financial support for: a Lifeline program, a state high cost fund, telecommunications relay service (TRS), and a statewide E-911 system to be completed by July 1997. According to Goldstein and Gooding (1995), as of December 1995, 18 states had approved (and eight states had pending) definitions of basic telephone services, which included these elements: single party, voice grade, touch tone, line with access to emergency services, directory assistance, operator services, long-distance services, and a white page listing. Eight states included TRS for the hearing-impaired, six states included a modem capable line, six

[5]As of May 1996, according to Rosenberg and Stanford (1996), the 13 states using a surcharge on telecommunications revenues to finance universal service support were Arizona, California, Colorado, Connecticut, Georgia, Hawaii, Idaho, Nevada, Oklahoma, Pennsylvania, Utah, Vermont, and Wyoming; the six states using a line charge to fund universal service support were Arizona, Colorado, Idaho, Kansas, Maine, and Utah; the nine states using other charges were Arkansas, Illinois, Indiana, Kansas, Oregon, Rhode Island, Texas, Vermont, Washington (USF), and Wisconsin.

states included privacy protection, five states included access to repair services.[6]

The targeted state approach to universal service would consist of a definition of universal service no greater than that adopted by the FCC. It could also consist of rate plans related to the geographic service areas applied in most rural areas. Rural consumers value being able to call family and friends in the next town as part of basic service, but a disproportionate number of their toll calls are made to reach services only available in the nearest large town. If the old concept of value-of-service pricing is applied, then rural basic rates would be very low because the geographic service area is so limited. State commissions have experimented with alternative and expanded geographic service areas that enable rural customers to access nearby towns and villages as part of their basic rates and not as toll calls. By way of comparison, the value of service for urban consumers is very high because the basic rate provides access to so many services.

Within the past decade, the Colorado PUC ordered a plan for county-wide extended area service (EAS); the Georgia PUC ordered a county seat plan; the Louisiana Commission established a local option service that caps rates on toll calls made within 22 miles of a county calling area; and in 1990 the Washington Utilities and Transportation Commission "adopted rules to identify and expand local calling areas where customers must rely excessively on toll service to meet basic calling needs" (Parker, Hudson, Dillman, Strover, & Williams, 1992, p. 75).

The alternative approach to universal service would be a creation of state requirements that are broader than that articulated in federal law or implemented by the FCC. In recent years, the political debate has drawn attention to the information-haves and the information-have-nots. Many state officials may want advanced telecommunications services, including information services, to be made available to all consumers.

In this regard, one policy option is to encourage or require carriers to provide access to advanced services at cost. According to Conte (1995), the 1994 Wisconsin law defines advanced services as "generally accessible, but

[6]According to Goldstein and Gooding (1995), as of December 1995, 18 states had approved (and eight states had pending) definitions of basic service, which included the following common terms: single party (16), voice grade (18), touch tone (20), access to emergency services (23), directory assistance (16), operator services (14), long-distance services (17), and a white page listing (18). The 18 states with the approved definitions were California, Colorado, Connecticut, Delaware, Florida, Georgia, Louisiana, Massachusetts, Michigan, Missouri, Nevada, New Jersey, Oklahoma, Oregon, Pennsylvania, Tennessee, Vermont, and Wyoming; the eight states with pending definitions were Alaska, Arizona, Hawaii, New York, North Carolina, Ohio, West Virginia, and Wisconsin. Eight states included TRS for the hearing-impaired: Arizona, Louisiana, Ohio, Oregon, Tennessee, Vermont, West Virginia, and Wisconsin. Six states included privacy protection: California, Colorado, Connecticut, New Jersey, New York, and Ohio. Five states included access to repair services: Alaska, California, Colorado, New Jersey, and Ohio.

are not required to be provided to every home and are not subsidized as heavily as basic services" (p. 36).

Quite a different option is to include advanced services as part of a state's definition of universal service for all consumers. Advocates of this approach should be aware that the social cost of doing so may be extraordinarily high. According to the Telecommunications Industries Analysis Project (TIAP) at the University of Florida, the cost of creating such a switched, broadband network could be more than $400 billion (Conte, 1995). The question that state policymakers must ask is: What will the public be willing to pay to provide advanced telecommunications services to those who can not pay for them?

Consumer advocates have consistently opposed major infrastructure investment to provide services for which demand has not been demonstrated because the ratepayers will get stuck with paying higher rates. Similarly, consumer groups have opposed mandating advanced telecommunications services as part of a definition of universal service because they fear the ratepayers will be burdened with the cost of providing two-way video services, which are of dubious social value and for which market demand is unproven. Daniel Pearl, writing in *The Wall Street Journal* on January 14, 1994, quoted Bradley Stillman, legislative counsel of the Consumer Federation of America, "I don't want to be forced to pay for the interactive video games or movies on demand of my neighbor down the street." According to Pearl, "telephone companies promote a broad definition of universal service because it allows them to lay miles of fiber-optic wire at customers' expense. His group [CFA] has the support of the powerful American Association of Retired Persons, which wants to ensure that rates for basic telephone service aren't raised to subsidize more speculative services" (p. A12).

In an age in which society's appetite for information seems to grow as fast, if not faster, than the speed at which the communications technologies are improved to disseminate it, the definition of universal service may be expanded from the notion of basic telephone services. Section 706(c) of the federal law defines advanced telecommunications capability as "high-speed, switched, broadband telecommunications capability that enables users to originate and receive high-quality voice, data, graphics, and video telecommunications using any technology."

In the future, will federal and state governments amend the definition of universal telecommunications services to include these broadband capabilities? Would ratepayers support surcharges on their telephone bills to pay for broadband services that others may want? Will the industry make huge investments to build broadband capacity when the demand for these services remains speculative? (Egan, 1996; Noll, 1997).

Society's support of public education may provide a useful analogy in answering these questions as to universal telecommunications services.

The public has demonstrated a general willingness to support public education for all children through high school. This is generally accepted by the public as a merit good, which benefits individuals and society as well. The public sector provides much less support for higher education—directly to individuals and indirectly through public universities—perhaps because of the belief that the primary beneficiaries (the students themselves and their parents) should pay more for those expenses. Given the social consensus on support for education, one might expect society to be willing to pay for universal access to basic telephone services—yet those who want and benefit from using the advanced telecommunications services should pay for those services themselves (Stoll, 1995).

STATE VIRTUAL VOUCHERS

NECA, on behalf of the FCC, distributes approximately $737 million through the current Universal Service Fund to keep the cost of basic services affordable for customers living in high-cost areas. Cost studies for defined areas determine the eligibility for these funds. For some time, economists have criticized the idea of providing direct subsidies to the companies serving these high-cost areas because it provides an insufficient incentive for them to improve operations and lower costs. Mueller (1995) asked, "But as long as the rural telecos are collecting universal service subsidies, why should anyone bother to develop and deploy radically different, more efficient ways to serve these areas?" (p. 228). Bill Frezza, president of Wireless Computing Associates, was quoted in *Communications Week* on November 27, 1995 (Goldstein & Gooding, 1995), making a similar criticism of current policy, "Has anyone asked whether there is a more direct way to help procure services on the open market? Food is more important than phone calls, but we sure don't ship food stamps directly to Stop-and-Shop and Grand Union based on some weird geo-political formula of hard-to-feed locations. Yet that's exactly what we do in the telecom business" (p. 11).

A virtual voucher would provide compensation to any carrier that provides services and assumes the obligations of providing service throughout the service area to any customer in a high cost area. The amount of the voucher would be calculated using a methodology very similar to current small area cost studies. This voucher would go directly to the carrier chosen by the customer and is valued at the difference between the actual cost of providing service and an average price set by the state commission. For example, a customer living in a high-cost area, where the actual cost of basic service is $50, would choose any carrier; if the average monthly cost of basic service in the state was $25, then the chosen carrier would receive the difference ($25) as a voucher from the state universal fund.

In California, the state commission is considering a virtual voucher approach. In its December 1995 report to the state legislature on maintaining universal service in competitive local phone market (CPUC Report, 1995), the commission explained current state policy as well as the rationale for adopting a high cost voucher fund:

> In areas that are expensive to serve, rates have been kept low in three ways. First, the California High-Cost Fund (CHCF) reduces rates for customers of small telephone companies by providing funds to some rural high-cost companies. All telephone customers currently contribute to the CHCF through a small charge on their monthly telephone bill. Second, by requiring large telephone companies to charge a single rate for basic service within their territory, rates are kept at reasonable levels in high-cost areas. Third, local exchange carriers charge access and toll rates which are priced above cost; these revenues may enable basic service to be priced below cost. These three mechanisms are only available to existing local exchange carriers.
>
> In a competitive environment, prices for services which provide revenue, like toll, will gradually be driven down. In addition, competitors will enter high revenue, low cost areas, putting pressure on large carriers to offer different rates in different areas. With local exchange competition, the Commission must develop a funding mechanism which targets high-cost areas throughout California, not specific telephone companies. This funding should be competitively neutral and available to all companies that provide basic service with the high-cost area. (pp. 13–14)

There are substantial advantages, at least in theory, to the idea of using a virtual voucher system instead of direct subsidies to maintain small companies in high-cost areas. The California PUC and other states face a major challenge in their efforts to design and implement such a system. The voucher idea also may attract prospective carriers that want to provide "one-stop shopping" (i.e., one bill for all telephone services including long-distance). If disenchantment with current direct subsidy programs to supplement the federal high-cost fund grows—as state costs escalate—governors and state legislators will give this innovative approach to ensuring universal service greater attention. State vouchers also have the distinct advantage of providing greater consumer choice for subscribers in high-cost areas.

Therefore, it is likely that a virtual voucher approach will draw increasing attention from state policymakers.

THE STATE DEBATES ON UNIVERSAL SERVICE

The top executives of the LECs have had long careers in the regulated environment in which the ethic of public service was strong. Hence, one might assume that they are deeply and sincerely committed to ensuring that the tradition of universal service is maintained in the competitive regime. Yet, as they advocate for universal service funds, testify about the size and scope of universal service mechanisms, present cost studies and proxy models, those executives are also accountable to stockholders of their companies. They are not disinterested parties in subsequent state debates on universal service. Furthermore, if the regulators require private companies to accept public obligations, such as COLR, the corporate leaders must insist that their companies be paid adequate compensation for them.

Similarly, the new entrants are acutely concerned about the size, scope, and beneficiaries of subsequent public policy on universal service. As providers, they are concerned about surcharges levied against their revenues to fund universal service programs, which might have the effect of providing distinct advantages to their competitors. State regulators will be thrust in the middle of these contentious debates, which have significant implications to the security and reliability of the public-switched network, the quality of services, the price and availability of services, and the modern notion of ensuring access to basic telephone services for everyone.

State elected officials may not be successful in deferring to the expertise of the state regulators to resolve the prominent policy conflicts. Unlike the past rate cases, state regulators may not have the opportunity to implement universal service programs without extensive public debates—the issue affects too many diverse corporate interests and competing social values.

For example, in 1997, US West and AT&T—opponents on proposed state legislation to change regulatory policy in North Dakota—launched major public relations campaigns, including television advertising, to influence public opinion. "Your phone bill will double in 4 years if this legislation passes," argued the opponents of the measure in newspaper advertisements. The substantive issues addressed by the proposed legislation were obscure to the public (it included a state universal service fund, but that provision was not controversial). Nevertheless, a sparsely populated state with a rather modest market for telephony services was subjected to spirited advocacy campaigns—a war of heated rhetoric—that advanced their competing corporate interests. This is not an isolated event (Bonnett, 1996).

Allen Hammond, a Professor at the University of Santa Clara Law School, is quoted in *Governing* (Conte, 1995) as saying: "The universal service debate is transplanting the debate over health and welfare to telecommunications and it has a good chance of becoming just as fractious" (p. 36). Consider these anecdotes, as reported by Conte (1995, 35–37):

1. Many telecommunication firms support the idea of having explicit fees that are paid by all service providers to a universal service fund, but there is great debate over who should get these subsidies and why. Andrew Lipman of MFS Communications, a competitive access provider in several large cities, offers this perspective: "No other commodity or service offered in the marketplace is sold at subsidized

prices based solely on geography—even those that are most essential to human life, such as food, health care, electricity, or gasoline."

2. Despite the trend of creating state universal service funds, the public sometimes rejects explicit subsidies. A case in point: in 1991 the Illinois state legislature rescinded a $0.15 monthly surcharge approved by the Illinois Commerce Commission to help reduce the cost of serving low-income families.

3. Universal service advocates are committed to keeping local basic rates low and argue that rate increases will prevent low-income households from maintaining phone service. Yet, in a counterintuitive case, Massachusetts regulators approved a local rate increase from $2.50 per month to $10.00 per month (and lower intrastate toll rates), with the chair of the Department of Public Utilities claiming that "the effect on universal service has been nil."

These anecdotes frame the debate for public policy. Universal service is an abstract social goal. Any method of achieving that goal requires social resources. Is the social goal sufficiently valued by the public to support a redistribution of resources? What value does society place on extending access to basic services to everyone? Are the ratepayers willing to support surcharges on their phone bills to provide direct subsidies to achieve universal service objectives? Which social values will prevail in these debates? How will the many loud voices of the competing corporate interests shape the public discourse and affect outcomes?

SUMMARY

The transition from the old regime of regulated monopoly providers with distorted rate structures toward the brave, new world of competition in local telephony will be problematic for all concerned. The traditions and legacies of the past will be difficult to shed. In addition, controversy abounds from every federal or state action. The local telephone market generates approximately $100 billion annually. That revenue will entice new entrants into these local exchange markets. With new competition will come new policy challenges for federal and state regulators, new market opportunities for the industry of providers, and presumably more choices by consumers.

Universal service has had a curious past: first, to promote industry consolidation and, then, in the 1970s and 1980s, as well-polished armor to defend the status quo of the monopoly providers. Shaped profoundly by this corporate advocacy, universal service is now commonly viewed as extending access to basic telephone services to everyone, regardless of income and regardless of geography, and serves as an important symbol for the egalitarian aspirations of our society. Extending access to basic services also provides benefits, albeit difficult to measure, for every other user of the publicly switched telephone network.

Advocates of universal service won major battles with the enactment of the 1996 Telecommunications Act. Those battles do not yet constitute a victory. Past accomplishments will have to be sustained in Congress, if voters threaten to rebel at the FCC's order, complain at the size of the high-cost fund to keep services affordable to rural users, ridicule the educational value of wiring schools to use the Internet, and object to the redistribution nature of these federal programs. Advocates must be concerned whether the FCC order will be sustained in the courts and in the realm of public opinion. And, finally, advocates will need to sustain their persuasive arguments in the states to achieve lasting victories. These will be fascinating developments to witness in the coming years.

REFERENCES

Bernt, P. (1996). *The eligible telecommunications carrier: A strategy for expanding universal service.* Columbus, OH: National Regulatory Research Institute.

Bonnett, T. W. (1996). *TELEWARS in the states: Telecommunications issues in a new era of competition.* Washington, DC: Council of Governors' Policy Advisors.

Borrows, J. D., Bernt, P. A., & Lawton, R. W. (1994). *Universal service in the United States: Dimensions of the debate.* Columbus, OH: National Regulatory Research Institute.

Cohen, J. E. (1992). *The politics of telecommunications regulation: The states and the divestiture of AT&T.* Armonk, NY: M.E. Sharpe.

Conte, C. R. (1995). Reaching for the phone. Governing, 8(5), 32–37.

Cooper, M. (1996). *Universal service: A historical perspective and policies for the twenty-first century.* Washington, DC: Boston Foundation and Consumer Federation of America.

CPUC *Report to the Legislature on Maintaining Universal Service in Competitive Local Phone Market.* (1995). (R.95-01-020, I95-01-021, December 20). San Francisco, CA: California Public Utility Commission.

Crandall, R. W. (1991). *After the breakup: U.S. telecommunications in a more competitive era.* Washington, DC: Brookings Institution.

Decision 96-10-066. (October 25, 1996). *Rulemaking on the Commission's Own Motion into Universal Service and to Comply with the Mandates of Assembly Bill 3643.* San Francisco, CA: California Public Utilities Commission.

Egan, B. L. (1996). *Information superhighways revisited: The economics of multimedia.* Boston: Artech House.

Gabel, D. (1995a). Federalism: An historical perspective. In P. Teske (Ed.), *American regulatory federalism and telecommunications infrastructure* (pp. 19–31). Hillsdale, NJ: Lawrence Erlbaum Associates.

Gabel, D. (1995b). Pricing voice telephony services: Who is subsidizing whom?" *Telecommunications Policy, 19*(6),453–464.

Gabel, D. (1996a). *An assessment of universal service.* Tallahassee, FL: Office of Public Counsel for the State of Florida.

Gabel, D. (1996b). *Improving proxy cost models for use in funding universal service*. Columbus, OH: National Regulatory Research Institute.

Governor's Telecommunications Policy Coordination Task Force. (1996). *Building the road ahead: Telecommunications infrastructure in Washington State*. Olympia, WA: Department of Revenue.

Goldstein, M., & Gooding, R. Z. (1995). *Universal service to universal access: The paradigm shift in citizens' use of telecommunications*. Tempe, AZ: International Research Center.

Itkin, L., & McLaughlin-Krile, E. (1995). *State telecommunications reform legislation authorizing local competition enacted during 1994 and 1995*. Washington, DC: National Conference of State Legislatures.

Kirchhoff, H. (1997a, March 20). Pac Bell proposes $305M rate cut. *State Telephone Regulation Report*. Alexandria, VA: Telecom Publishing Group.

Kirchhoff, H. (1997b, March 20). Price cap regulation has become the norm in eastern states. *State Telephone Regulation Report*. Alexandria, VA: Telecom Publishing Group.

Leighton, W. (1995). *Telecommunications subsidies: Reach out and fund someone (whether you want to or not)*. Washington, DC: Citizens for a Sound Economy.

Mueller, M. (1995). Universal service as an appropriability problem: A new framework for analysis. In G. W. Brock (Ed.), *Toward a competitive telecommunications industry: Selected papers from the 1994 Telecommunications Policy Research Conference* (pp. 225–234). Mahwah, NJ: Lawrence Erlbaum Associates.

Noam, E. M. (1994a). Beyond liberalization: From the network of networks to the system of systems. In *Telecommunications Policy, 18*(4), 286–294.

Noam, E. M. (1994b). Beyond liberalization III: Reforming universal service. In *Telecommunications Policy, 18*(9), 687–704.

Noll, A. M. (1997). *Highway of dreams: A critical view along the information superhighway*. Mahwah, NJ: Lawrence Erlbaum Associates.

Parker, E. B., Hudson, H. E., Dillman, D. A., Strover, S., & Williams, F. (1992). *Electronic byways: State policies for rural development through telecommunications*. Boulder, CO: Westview Press.

Pearl, D. (1994, January 14). Debate over universal access rights will shape rules governing the future of communications. *The Wall Street Journal*, A12.

Rosenberg, E., & Stanford, J. D. (1996). *State universal service funding mechanisms: Results of the NRRI's survey*. Columbus, OH: National Regulatory Research Institute.

Stoll, C. (1995). *Silicon snake oil: Second thoughts on the information highway*. New York: Doubleday.

Stone, A. (1989). *Wrong number: The breakup of AT&T*. New York: Basic Books.

Teske, P. (1995). Introduction and overview. In P. Teske (Ed.), *American regulatory federalism and telecommunications infrastructure* (pp.3–17). Hillsdale, NJ: Lawrence Erlbaum Associates.

Vogelsang, I., & Mitchell, B. M. (1997). *Telecommunications competition: The last ten miles*. Cambridge, MA: MIT Press.

15

Breaking the Bottleneck and Sharing the Wealth: A Perspective on Universal Service Policy in an Era of Local Competition

Robert K. Lock, Jr.
Competitive Strategies, Chicago, IL

One of the most debated and least understood concepts in all of telecommunications policy is that of universal service. Over the past century universal service has been a dynamic concept, changing significantly depending on the environment. Contrary to popular belief, it was originally implemented in 1907 as a business strategy by AT&T's Theodore Vail, to promote connection to AT&T's network. In its modern incarnation, the ability to make and receive calls on the Public-Switched Telephone Network (PSTN) is viewed—if not as a basic human right, then—at least as a necessity for full participation in modern society (Mueller, 1993).

A review of the literature on the subject reveals that what at first appears to be a relatively simple idea can actually be a slippery and ideological concept, which has been used and manipulated by various special interests throughout the development of the telephone industry to support their own case for special treatment (Blackman, 1995). As always, in relation to concepts like universal service, "even in its most private deliberations, big business masks its self-interest behind the pretense of benevolence" (Von Auw, 1983, p. 224) From local exchange carriers (LECs), to interexchange carriers (IXC's), to competitive access providers (CAPs), regulators, and consumer groups, the concept of universal service has meant different things at different times depending on a host of different factors. Consequently, in every jurisdiction, from the Federal Communications Commission (FCC) to the various state regulatory commissions, there may be several very different definitions of the concept operating concurrently, resulting in uncertainty and confusion for all participants in the regulatory process.

Just as with the long-distance industry before it, local exchange has been beset by an influx of competitive entry at the fringe of what once were considered natural monopoly markets. The combination of technological innovation and the changing nature of demand for telecommunications has destroyed the old natural monopoly character of the telephone industry (Hunter, 1983). Technological change has caused production costs to fall, allowed new products and providers to emerge, and raised significant questions as to the ability of the current regulatory apparatus to deal with these changes in the market (Shin & Ying, 1992). As part of the transition to a competitive environment, it has become necessary for policymakers to reevaluate various assumptions that have been made regarding universal service.

Now that the traditional regulatory contract has been modified to allow competitive entry, competition is expected to supplant the protection previously afforded consumers by regulatory safeguards. The consumer, in addition to receiving the same level of protections as previously enjoyed under regulation, is promised increased innovation in telecommunications products and services. For the regulated carriers, the result of the contract modification will be the opportunity to enjoy freedom from oversight, increased flexibility to deal with an unpredictable environment, and whatever profits the competitive market will allow.

As the discussion illustrates, the disjointed evolution of the concept and the lack of a single definition of universal service makes the initial analysis of the concept difficult. This chapter advocates a multidisciplinary approach to the examination of the concept of universal service. In the process, it uses developments in Illinois over the last 10 years to illustrate a process by which rational universal service policies can be developed that are compatible with policies on local competition.

Ultimately, all participants should acknowledge that although the magnitude and pace of change may be significantly greater than in past eras, the regulatory process is sufficiently dynamic to accommodate that change. Recognizing the speed of change, as well as the quasiubiquitous control incumbent providers retain over their markets, regulators can expeditiously, yet judiciously examine the particular characteristics of the markets that they oversee and develop rational and flexible policies that match the needs of those markets.

REGULATORS NEED TO DEVELOP A SINGLE DEFINITION OF UNIVERSAL SERVICE

As regulators across the nation begin to confront the issues associated with universal service in an age of local competition, the first step toward the development of appropriate policies will be the construction of a working definition of the concept. Once regulators agree on a definition, they can begin the process of gathering and analyzing data on factors such as variation in consumer demand across the urban–rural continuum; the impacts of cultural, religious, social and economic factors on decisions not to subscribe; and other factors associated with the concept. Only after a definition has been agreed on can rational policies be constructed that acknowledge the unique characteristics of individual jurisdictions.

The definition itself should be fairly consistent across jurisdictions, however, each jurisdiction may vary as to the emphasis that it places on various aspects of the concept. The following four principles should govern regulators' deliberations during the development of policies on universal service:

1. In analyzing the universal service concept in an era of local competition, regulators need to move toward cost-based rates and nondiscriminatory access to network functionalities;
2. Regulators must take the time necessary to construct a record of evidence sufficient to allow for rigorous analysis and resolution of these complex issues;
3. Regulators must tailor policies to the characteristics of their individual jurisdictions; and,
4. During this process, regulators need to adopt flexible regulatory policies with monitoring mechanisms that will allow them to make changes in response to conditions in the markets in their particular jurisdictions.

Given the strong interrelationships between competition and universal service issues, the adoption of policies that deviate significantly from the four principles could introduce distortions into the local exchange market through the imposition of rigid policies that do not match the characteristics of the markets for which they were crafted. These distortions would be very difficult to rectify absent prolonged litigation or legislative action.

The Need for a Multidisciplinary Approach to Analysis and Reformation of Universal Service Policies

Despite its murky development, it is clear that universal service has become a cornerstone of telecommunications policy in the United States. As early as 1919, state regulatory commissions recognized that "[t]he use of the telephone is a business and social necessity, and the efforts of a great telephone company ... ought to be directed to some extent toward the idea of encouraging the use of the telephone rather than discouraging that use" (Horwitz, 1989, p. 135). Inherent in the concept is some vague presumption that it is in the public interest to seek maximum availability of an ever-increasing array of advanced communications services. According to a recent filing before the Illinois Commerce Commission, universal service is described by various parties as including dial tone, unlimited usage in the local calling area, touch-tone capability, single-line/single-party service, and public functions such as: 911, 411, TDD, annoyance call bureau, and white pages listing (*Illinois Bell Telephone Company Proposed Introduction of Ameritech's Customer's First Plan in Illinois,* ICC Docket No. 94-0096, consol.; Citizen's Utility Board Ex. 1; MCI Ex. 1.0). In its common construction, universal service is of sufficient importance to justify rate subsidies and other support mechanisms, legal obligations to serve, and a host of other regulatory obligations. Contemporary conceptualizations of universal service by consumer advocates appear intent on expanding the concept beyond plain old telephone service (POTS), and increasing penetration of the technology from the current 94.2% to as close to 100% as possible. For households with less than $10,000 annual income, the penetration rate may be as low as 79.8%, and for individual ethnic groups the level may be significantly lower still (National Telecommunications and Information Administration [NTIA], 1995).

The convergence of computing and communication technologies has injected a misplaced sense of urgency into the debate on the concept. Although some would prefer to view technological changes as irrelevant to the discussion, others have taken the position that the universal service concept should evolve as technology does. Recently, discussion of the inclusion of more advanced communications technologies in the definition has increased, as resources such as the Internet are increasingly viewed as entitlements. How is it that the telephone went from a feared intrusion into the sacrosanct parlors of Victorian America to a technology that is considered by some to be so vital that its placement in our homes has, at times, taken on the status of religious shrines (Marvin, 1988)? These issues will have to be addressed if we are to craft reasonable universal service policy. As always, the main question will be who will pay for it.

It is critical to undertake a multidisciplinary approach to the analysis of universal service, given that the concept is a veritable monument to the influences of a multitude of disciplines over the past 90 years. Law, economics, corporate strategy, as well as the social and political sciences, have all contributed to the patchwork evolution of the concept. Only by undertaking an analysis that examines principles from all of these relevant disciplines can a rational universal service policy, which reflects the characteristics of individual jurisdictions, be defined and constructed. Restricting analysis to less than the entire menu of perspectives risks the development of myopic and mismatched policies that could ultimately do more harm than good.

The Significance of the Communications Act of 1934

Most contemporary policy statements on universal service cite the Communications Act of 1934 as the foundation of the concept. According to advocates before both state and federal regulatory agencies, the universal availability of basic telecommunications service at affordable rates has been a fundamental element of telecommunications policy in the United States since the enactment of the Communications Act of 1934.

Since the 1970's, the Communications Act of 1934 has been cited regularly for the proposition that not only is access to the PSTN to be provided to all at reasonable rates, in some cases the public interest requires that that access be subsidized. Consequently, any examination of the modern definition of universal service should start there.

Section I, Title I, of the Communications Act of 1934 states that the goal of communications policy is "to make available, so far as possible, to all people of the United States, a rapid, efficient, nationwide, and worldwide wire and radio communication service with adequate facilities at reasonable charges."

Even a cursory review of the just cited language reveals that the concept of universal service is not specifically defined in the statute. A review of the legislative history surrounding the enactment in 1934 reveals that its sole purpose was actually to consolidate the functions of the Interstate Commerce Commission and the Federal Radio Commission into the newly formed Federal Communications Commission. In fact, the legislation changed no existing laws and had nothing specifically to do with the modern concept of universal service (Horwitz, 1989).

Contrary to modern conceptions of the origins of universal service, research suggests that the concept was not linked to the Communications Act of 1934. Instead, the concept evolved as a business strategy developed by AT&T that was shrewdly packaged in public interest rhetoric (Mueller, 1993). Any definition that is adopted must be cognizant of this fact, as well as Parker's (1989) recognition that "Universal service has never implied an entitlement program under which U.S. residents would have a right to telephone service at government expense. Rather, the goal ... is to ensure that the structure of the industry makes telephone service universally accessible and affordable" (p. 12).

Once regulators recognize the origins of the concept, it will be easier to modify whatever misconceptions they might have been operating under and move forward toward a definition that reflects the needs of their jurisdictions.

REGULATORS NEED TO CRAFT UNIVERSAL SERVICE POLICIES THAT REFLECT THE CHARACTERISTICS OF THEIR INDIVIDUAL JURISDICTIONS

Universal service is a dynamic concept that should be viewed within the context of a jurisdiction's stage of economic development, as well as within the context of relevant economic, social, and political objectives (Blackman, 1995). In this environment, it is the role of regulators to balance the incentives of the parties and exercise patience and good judgment, taking care to develop a sound record of evidence on which to construct a transitional regulatory framework.

During this tumultuous period, an incredible strain is being placed upon the resources of all participants in the regulatory process. LECs—both incumbents and new entrants—cable television companies (CATV), cellular carriers, and regulatory agencies are expending vast amounts of resources in attempts to resolve some of the most complex issues that they have ever confronted. Because of intense economic and political pressures being applied in most jurisdictions, many of these issues are being handled on an expedited basis. The demands of a multitude of interrelated dockets—each as critical to the overall outcome as the next—creates a very difficult and dangerous situation. In such an environment, the potential for the creation of mismatched or inappropriate policies is increased exponentially.

It is critical that all parties realize that there are no shortcuts to this process. The process is an imperfect one and adjustments will have to be made. Regulators should not expect to come up with the policy equivalent of the great American novel. Nor should they expect to adopt whole-cloth the policies of other jurisdictions. Instead, they must craft policies suited to the dynamics of the markets in their jurisdictions, allowing room for adjustment as the conditions in those markets change.

I am troubled to a certain extent by the number of states that have very recently introduced policies on local competition issues. Not because of the fact that they have done so, but because of the possibility of a bandwagon effect. As Abrahamson and Rosenkopf (1993) explained, bandwagons are diffusion processes whereby organizations adopt an innovation, not because of their individual assessments of the innovation's efficiency or returns, but because of the pressure caused by the sheer number of organizations that have already adopted the innovation. The speed with which states have adopted policies suggests that, rather than developing an evidentiary record and comparing alternatives before making a choice, decision makers in some states may be simply imitating the choices made by others. Although imitation in some cases may be a cost-effective way of developing policy, such conduct may be dangerous in terms of the long-term viability of a competitive local exchange market and the development of rational and innovative approaches to universal service policy (Pingle, 1995).

The pace of regulatory change in some jurisdictions suggests that perhaps some regulators are adopting whole-cloth the policies of early adopters of regulatory reform without regard to the needs of their individual jurisdictions. There is nothing wrong with using components from other jurisdictions' policies. A wise person learns from the failures, as well as the successes of others. However, regulators should guard against imitative behavior–adopting policies out of fear of losing legitimacy among their peers or other such motivations–rather than responding to the exigencies of their particular circumstances.

The fact is, competition at the local exchange level is nascent at best in those jurisdictions that have worked aggressively over the last 10 years to develop it. Many jurisdictions, particularly those with a large proportion of rural subscribers, are not likely to realize effective competition for quite some time, if ever. The dynamics of competing carriers negotiating contentious interconnection and compensation agreements, not to mention the ability of the incumbent LECs to effectively dictate the pace of the process, provides sufficient time to develop policies that reflect the individual needs of the jurisdiction to which they were meant to apply.

Because the ramifications of mistaken judgments based on rushed and haphazard analysis are so great, it is incumbent on regulators to control the pace of regulatory change. Rational policy based on evidence, and not supposition or some perceived need for legitimacy, must guide regulators through this complex maze of issues.

THE NEED FOR REGULATORS TO ADOPT POLICIES PROMOTING COST-BASED RATES AND THE ELIMINATION OF BARRIERS TO ENTRY: A LOOK AT ILLINOIS

Once a jurisdiction has opted to promote the development of a competitive telecommunications marketplace, it is imperative that the policies that are adopted send the appropriate signals to all participants in that marketplace. Consumers, service providers, manufacturers, and regulators must all share the same understanding of the costs of service and who should bear those costs. Subsidies (both explicit and implicit) developed in a monopoly environment must be reexamined based on the dynamics of a competitive market.

To eliminate antiquated subsidies while also ensuring that marginal customers, such as low-income and rural subscribers, receive the benefits of advancing telecommunications technology, regulators should implement policies that promote: (a) cost-based rates for telecommunications services, (b) the development of other appropriate preconditions for the establishment of a truly competitive local exchange marketplace, and (c) the establishment of targeted subsidy mechanisms funded in a competitively neutral manner. Only in this way will it be possible for the uneconomic customer of today to become a new competitor's economic customer tomorrow (Blackman, 1995).

Regulators in Illinois have taken significant steps in addressing points (a) and (b), and are in the process of addressing point (c) in a comprehensive examination of universal service issues. The steps that have been taken thus far have proven successful in attracting new entrants to the Illinois local exchange, and should prove useful to those only recently embarking on this journey.

Movement Toward Cost-Based Rates

By far, the most important of the initial actions taken by the Illinois Commerce Commission (ICC) after the divestiture of AT&T was the development of a regulatory mechanism that allowed for the gradual movement toward cost-based rates and the elimination or reduction of many historical subsidies. Unlike other jurisdictions, Illinois classified central office equipment (COE) as nontraffic sensitive (NTS) and addressed it in a framework that provided for its eventual phase out.

The effect of this action was to remove some existing cross subsidies between companies and to transfer the recovery of NTS costs from intrastate toll costs to customers of exchange companies. During the same period, the ICC also established an intrastate high-cost fund (HCF). The purpose of the HCF was to mitigate the impact that the complete phase down of intrastate NTS costs from IXC common line charges had on LECs. The HCF is funded by IXCs and primary toll carriers on an allocated, for example, flat-rate basis as determined by each carrier's projected intrastate toll minutes of use.

In addition, in 1986 the ICC authorized the establishment and implementation of the Primary Toll Carrier (PTC) concept as a transitional method for deaveraging intra Metropolitan Statistical Area (intraMSA) toll rates (ICC, 1986). Under the PTC, the LEC with the most significant toll presence in an MSA was designated the PTC and, for that MSA, responsible for developing and filing toll rates, developing and administering a compensation plan for settling toll traffic accounts among carriers, and coordinating the toll network. The PTC plan was terminated by stipulation and agreement of the parties on December 20, 1995, pursuant to the ICC's order in the Customers First proceeding, which required that staff investigate and file proposed rules for the elimination of the PTC plan within 6 months from the entry of the order (Commission's Order, 1995).

The result of all of these actions has been the realization of intrastate toll and access rates being driven toward cost. Ultimately, this has positioned Illinois as a prime location for the development of a vibrant competitive local exchange environment, as end users and competitors alike are sent appropriate price signals for telecommunications services. This has resulted in all carriers having a reasonable opportunity to compete for all customers in the local exchange, rather than just the high-volume business market. As politically difficult as these moves may have been, it is difficult to believe that an effectively competitive market could develop without them.

Elimination of Legal Barriers to Entry Into the Local Exchange

In addition to pursuing policies designed to move toward cost-based rates, Illinois has also moved deliberately toward establishing an environment conducive to the development of competition in the local exchange. The catalyst for the actions of the ICC over the past 10 years in the area of local competition was the adoption of the Universal Telephone Service Protection Law of 1985 (UTSPL). This law, embodied in Article XIII as an amendment of the Illinois Public Utilities Act, eliminates exclusive certificates of service, and allows for the introduction of competition "when consistent with the protection of consumers." At the time of its adoption, UTSPL was considered extremely progressive, and has served as a model for a number of other states' regulatory policies in telecommunications.

Under the regulatory decisions discussed earlier, and the guidance of the UTSPL, regulators in Illinois have established an environment that has proven extremely conducive to competitive entry—as witnessed by the granting of certificates of facilities-based local exchange service authority to carriers such as TCG, MFS, Midwest Fibernet and MCI Metro, as well as AT&T, LCI, and U.S. Network.

However, although the presence of new entrants acknowledges the elimination of a legal barrier to entry, it is not sufficient in and of itself to establish full and effective competition in the local exchange. For this, it is necessary to establish a regulatory environment that sends the proper signals to all market participants. In Illinois, the consolidated dockets that came to be known as the Customers First case (Commission's Order, 1995) have provided the vehicle for the ICC to take significant steps in the development of an effectively competitive market for telecommunications services.

The Customers First case involved four consolidated dockets and two companion rulemakings dealing with issues associated with the development of competition in the local exchange. Ameritech Illinois' request to unbundle certain noncompetitive portions of its local exchange for resale to new entrants was linked with its requested waiver of the Modified Final Judgment (*U.S. v. Western Electric*, 1982) for a trial of interLATA services in the Chicago area that was filed with the U.S. District Court for the District of Columbia. In addition, AT&T filed a petition for the investigation and establishment of conditions necessary for the establishment of effective local exchange competition. The rulemakings involved rules on presubscription, and unbundling and interconnection. Final Orders were entered by the ICC on April 7, 1995 (ICC, 1995).

Among the issues that were addressed in the Customers First case and associated rulemakings were interconnection, unbundling, compensation for traffic termination, and presubscription. The following issues were deferred for further inquiry: numbering issues (portability, administration of numbering resources); resale, reform of outdated regulatory policies (regulatory treatment of new entrants); access to pole attachments, conduits, rights-of-way and other operational issues; and universal service. The following discussion provides some insight on the present status of ICC examination of these issues.

Elimination of Operational Barriers to Entry

Although some have construed the Customers First case as a comprehensive and final template for resolution of policy issues related to competition in the local market, the truth is that it is really more of an evolving regulatory framework. The process has been slower than either the incumbents or the new LECs would like. However, in the ultimate analysis, none would prefer the alternative—rigid and inflexible regulatory policies based on insufficient evidence. With the Customers First Orders, Illinois has endeavored to examine the universal service issue as one very critical aspect of a blueprint for the development of a competitive market in the local exchange. In that case, the ICC undertook an aggressive, but measured approach to regulatory policy development. Some issues were decided that related specifically to Ameritech, whereas others were decided generically through rulemakings. Still other issues for which the ICC lacked sufficient evidence were prioritized for future examination. Universal service is one of these.

As part of the mandate of the orders in Customers First, the ICC undertook a comprehensive examination of universal service issues. A workshop process was conducted that established five working groups to gather and assess information on the following issues:

1. Why people are falling off of the network;
2. Nonfunded assistance mechanisms, such as toll blocking;
3. Funding of explicit low-income assistance programs, such as Lifeline and Link-Up;
4. Comprehensive treatment of issues related to serving low-income and high-cost customers; and,
5. Issues related to administration of the HCF.

Under the terms of the ICC's Order in Customers First, the Staff filed a report on March 28,1996 that outlined the issues that arose in the workshop process and recommended steps for future examination of various universal service issues.

Other Regulatory Developments

Other ICC actions intended to facilitate the development of competition for telecommunications services were the adoption of rules requiring imputation tests for switched interexchange services and local competitive services. These rules were adopted to protect competitors by attempting to ensure that the companies providing noncompetitive inputs to competitors charge them the same rates that those companies explicitly or implicitly charge themselves. In addition, the ICC has adopted rules requiring that carriers providing both competitive and noncompetitive services demonstrate that the services are priced above long run service incremental costs (LRSIC), as a protection against cross subsidy.

Also, under the terms of the orders in Customers First, the ICC Staff has been charged with examining existing telecommunications rules and requirements, and their application to new LECs. In this workshop process, parties are conducting a comprehensive review of all current regulations facing LECs, with the intent of (a) imposing regulatory requirements on carriers that match the characteristics of the individual carriers and the market, and, (b) eliminating those regulatory requirements that are incompatible with the new competitive environment.

Although this process has been both time- and labor-intensive, the ICC has nonetheless worked expeditiously to provide a regulatory framework that encourages cost-based rates and competitive entry, while taking into account the specific characteristics of the market for telecommunications services in Illinois. In so doing, Illinois has become one of only a few states that have taken steps to invite competition for all customers in all markets, not just high-volume business customers. This is the key to a successful universal service policy in an era of local competition.

CONCLUSION

The concept of universal service has been a central focus of regulation since the turn of the century. It has evolved as an integral part of the regulatory environment for telecommunications in the United States. Recent research demonstrates that the contemporary conceptualization of universal service is actually an extension of the original corporate mission of AT&T: to get as many people on their network as possible, and thereby justify a regulated monopoly for the provision of telephone service (Mueller, 1993). The development of the concept tracks the evolution of AT&T, changing to conform to contemporary public interest rhetoric.

The terms of the traditional regulatory contract between AT&T and its regulators at the state and federal levels allowed for the construction of a state-of-the-art telecommunications network that has reached upward of 94% of the nation's households. In order to accomplish this, AT&T, independent providers, and regulators constructed a set of cross-subsidy mechanisms that allowed providers to keep prices low for high-cost rural and low-income residential customers. Under this paradigm, long-distance rates have generally subsidized local rates in order to promote public policies such as universal service (Weinhaus & Oettinger,1988).

The effects of this contract have been to stimulate investment to serve customers who may otherwise not have been served through the operation of the competitive market. The economic justification for regulation of telecommunications was based on significant economies of scale in local exchange and long-distance markets, as well as the expense and inconvenience to consumers of having to deal with parallel competing networks (Phillips, 1988).

Radical changes in the telecommunications industry in recent years have caused all parties in the telephone debates to reexamine the concept of universal service and its appropriateness in the present environment. In the transition to a competitive industry, regulators must take deliberate and measured steps in the development of local competition and universal service policies. Information must be gathered and analyzed in a timely manner to ensure that policy decisions are based on the most current information. At a minimum, baseline assessments of penetration levels must be evaluated and preliminary and long-term goals established for the targeting of groups that consistently are un- or underserved. Consistent monitoring of service quality will be critical during this period to ensure that existing customers do not suffer as competitors seek to expand their customer bases or to protect their shares of the market.

In addition to pushing for cost-based rates for telecommunications services, states should develop policies that acknowledge the natural tendencies of market participants and are tailored to the specific circumstances of their jurisdictions. Particular attention should be focused on issues such as consumer demand for telecommunications services, the ability of the existing infrastructure to satisfy that demand (and any reasonable projections of demand growth), and the characteristics of customer groups that fall below a reasonable level of penetration.

Most importantly, regulators must take great pains to control the pace of the process and manage overburdened regulatory resources. Despite the presence of competition in select markets, regulators continue to have a statutory obligation to protect both consumers and those entities that they regulate. In the end, the development of flexible regulatory policies based on rational principles and cost-based rates will allow for a smooth transition to competition and the realization of consumer-oriented universal service goals that evolve as the industry advances.

REFERENCES

Abrahamson, E., & Rosenkopf, L. (1993). Institutional and competitive bandwagons: Using mathematical modeling as a tool to explore innovation diffusion. *Academy of Management Review, 18*, 487–510.

Blackman, C. (1995). Universal service: Obligation or opportunity? *Telecommunications Policy, 19*, 171–176.

Commission's Order. (1995, April 7). *Illinois Bell Telephone Company Proposed Introduction of Ameritech's Customer's First Plan in Illinois*, ICC Docket No. 94-0096 consol.

Horwitz, D. (1989). *The irony of regulatory reform*. New York: Oxford University Press.

Hunter, H. (1983) Pricing telephone service in the 1980's. In A. L. Danielson & D. R. Kamerschen (Eds.), *Current issues in public-utility economics: Essays in honor of James C. Bonbright* (pp. 49–62). Lexington, MA: Lexington Books, D.C. Heath & Co.

Illinois Commerce Commission (ICC). (1995, April 7). *Illinois Commerce Commission On Its Own Motion, Adoption of Rules Relating to intra-Market Service Area Presubscription and Changes in Dialing Arrangements Related to the Implementation of Presubscription* ICC Docket No. 96-0048; *Illinois Commerce Commission On Its Own Motion, Adoption of Rules on Line-side Interconnection and Reciprocal Interconnection* ICC Docket No 96-0049.

Marvin, C. (1988). *When old technologies were new: Thinking about electronic communication in the late nineteenth century*. New York: Oxford University Press.

Mueller, M. (1993). Universal service in telephone history: A reconstruction. *Telecommunications Policy, 17*, 352–369.

National Telecommunications and Information Administration (NTIA). (1995). *Falling Through the Net: A Survey of the "Have Nots" in Rural and Urban America*. Boulder, CO: U.S. Department of Commerce, Institute for Telecommunications.

Parker, E. (1989). *Rural America in the information age: Telecommunications policy for rural development*. Lanham, MD: University Press of America.

Phillips, C. (1988). *The regulation of public utilities: Theory and practice*. Arlington, VA: Public Utilities Reports, Inc.

Pingle, M. (1995). Imitation versus rationality: An experimental perspective on decision making. *Journal of Socio-Economics, 24*, 281–315.

Shin, S., & Ying, J. (1992). Unnatural monopolies in local telephones. *RAND Journal of Economics, 23*, 171–183.

Von Auw, A. (1983). *Heritage & destiny: Reflections on the Bell System in transition*. New York: Praeger.

U.S. v. Western Electric. (1982). U.S. Dist. Ct., Dist. of Columbia, 552, F.SUPP.131

Weinhaus, C., & Oettinger, A. (1988). *Behind the telephone debates*. Norwood, NJ: Ablex.

Author Index

A

Abrahamson, E., 242, *248*
Access Charge Reform, 181, 185, *187*
Adams, S., 45, *57*, 114, 118, 120, 121, 122, 131, *133*, *134*
Adams, W., 28, *36*
Administrative Procedures Act, 186, *187*
Aikman, W. F., 22, 23, 23n4, 23n5, 26, *36*
Alba, R. D., 27, *36*
Albright, H., 45, *57*, 114, 115, 116, 117, 118, 119, 120, 121, 122, 123, 124, 125, 126, 127, 128, 129, 130n4, 130n5, 131, *133*, *134*
Alexander, K., 26, *35*
Aley, J., 142, 143, *155*
Allen/Fish, 151, *155*
Alltel v. FCC, 185, *187*
Amendment of Part 36 of the Commission's Rules and Establishment of a Joint Board, 180, *187*
And a Big Loss at AT&T, 141n6, *156*
Anderson, R. H., 62, *65*, 144n7, 145n9, *156*
Annual Report of the AT&T Company, 147, *156*
Appropriations for 1997: Hearings Before the Subcom. on Labor, Health and Human Ser., and Ed., and Related Agencies of the House Comm. on Appropriations, 100, *107*
Arboleda, J., 27, *35*
Arellano, M., 102, *107*
Arnst, C., 101, *107*
Aron, D., 168, *177*
Artzt, E. L., 102, *107*
Aspden, P., 193, *210*
Astin, A. W., 27, *37*
Astin, H. S., 27, *37*
AT&T, 189, *209*
AT&T Corporate Archive, 1900, 149, 149n11, 150, *156*
AT&T Corporate Archive, 1909, 148nf, *156*
AT&T Corporate Archive, 1880-1908, 150, 151, *155*
Ausubel, L. M., 141n5, *156*

B

Beilock, R., 140, *156*
Belinfante, A., 189, 190, 191, *209*, *211*
Bell Atlantic Telephone Companies v. FCC, 55n9, *57*
Berg, S., 115, 116, 117, 118, 119, 123, 124, 125, 126, 127, 128, 129, 130n4, 130n5, *133*, *134*
Bernt, P., 221, 224, 225, 227, *234*
Bernt, P. A., 215, *234*
Bethel, N. U., 154, *156*
Bikson, T. K., 62, *65*, 144n7, 145n9, *156*
Birenbaum, W. M., 27, *35*
Black, M., 16, *35*
Blackman, C., 237, 241, 243, *248*
Board of Regents v. Roth, 74n13, *82*
Boddie v. Connecticut, 74n13, *82*
Bolter, W., 189, *209*
Bond, D., 45, *57*, 120, 121, 122, 131, *133*
Bonnett, T. W., 226, 232, *234*
Borenstein, S., 140, *157*
Borrows, J. D., 215, *234*
Bosley, J., 115, 116, 117, 118, 123, 124, 125, 126, 129, 130n4, 130n5, *133*, *134*
Bowles, F., 26, 27, 33, *35*
Bradsher, I., 106, *107*
Bray, H., 102, *107*
Brock, G., 20, *35*
Brooks, J., 29, *35*
Brown, J. S., 60, 61, *65*
Brown, K., 199, *209*

Brown, R., 192, *211*
Brown, R. H., 15, *35*
Browning, J., 70, *82*
Browning, J. S., 60, *65*

C

Cabinet, 200, 207, *209*
Calaway, G., 45, *57*, 114, 117, 118, 120, 121, 122, 131, *133*, *134*
Campbell, C. D., 105, *107*
Cantwell v. Connecticut, 60, *65*, 71, 74n13, *82*
The Card: A work in progress, 141, *157*
Carlini, J., 103, *107*
Charlton, D., 115, 116, 119, 123, 124, 125, 126, 127, 128, 129, 130n4, 130n5, *134*
Cherry, B., 47, 50n2, *57*
Chesapeake & Potomac Telephone Company, 77, 78, *82*
City of Cincinnati v. Discovery Networks, 73, *82*
Clark, A., 114, 115, 117, 123, 129, 130n4, 130n5, *133*, *134*
Clinton, W., 192, 200, *209*
Cohen, J. E., 215, *234*
Commission's Order, 244, 245, *248*
Common Carrier Bureau, 131, *133*
Communications Act of 1934, 70, 80, *83*, 101, *107*, 179, 180, *187*
Communications Act of 1936, 70
Competition at the Local Loop: Policies and Implications, 102, *107*
The Conference Board, 120n3, *133*
Conference Report for the Telecommunications Act of 1966, 135, *156*
Connors, K., 45, *57*, 114, 118, *134*
Conte, C. R., 218, 228, 229, 232, *234*
Cooper, M., 221, *234*
Copeland, P., 45, *57*, 114, 115, 116, 117, 118, 119, 120, 121, 122, 123, 124, 125, 126, 127, 128, 129, 130n4, 130n5, 131, *133*, *134*
Cowen, T., 63, *65*
Cowles, R., 119, 127, 128, *134*
C&P Telephone Company, 191, *209*, *210*
CPUC Report to the Legislature on Maintaining Universal Service in Competitive Local Phone Market, 231, *234*
Crandall, R. W., 216, 217, *234*
Crandall v. Nevada, 73n8, *83*

D

Davis, R., 101, *107*
de Sola Pool, I., 61, *65*, 151, *157*
Department of Telecommunications and Energy, City of New York, 77, *83*
Dillman, D. A., 32, *37*, 228, *235*
Duguid, P., 60, 61, *65*
Dunbar, J., 45, *57*, 115, 116, 117, 119, 120, 121, 122, 123, 124, 125, 126, 127, 128, 131, *133*, *134*
Dupont, D., 129, 130n4, 130n5, *134*

E

Educational Policies Commission, 27, *35*
Educational Tech.: Hearing Before the Subcomm. on Labor, Health and Human Services, and Education, and Related Agencies of the Sen. Comm. on Appropriations, 99-100, *107*
Edwards, N., 26, *35*
Egan, B. L., 219, 229, *234*
Ehrenhalt, S., 100, *107*
el-Khawas, E. H., 27, *37*
Emerson, T. I., 60, *65*, 71, 72, *83*

F

Fagen, M. D., 30, *35*
FCC Report and Order, 169, 170, 171, *177*, 201, 202
Federal Communications Commission, 75, *83*, 147, 149, *156*, 189, *210*
Federal-State Joint Board on Universal Service, 180, *187*, 201
Federal-State Joint Board Staff, 75, 75n15, 76, 76n16, 76n17, 76n18, *83*
Field Research Corporation, 77, 78, *83*, 191, *210*
Fischel, W. A., 105, *107*
Fischer, C., 29, 30, *35*, 147, 151, *156*
Fish & Wallace, 152, *156*
Fish & Wheeler, 153, *156*
Fish & Yost, 152, *156*
Fiumara, G. C., 15, *35*
Ford, W. S., 153, 154, *156*

G

Gabel, D., 151, *156*, 218, 221, 227, *234*
Garbanati, L., 45, *57*, 114, 115, 116, 117, 118, 119, 120, 121, 122, 123, 124, 125, 126, 127, 128, 129, 130n4, 130n5, 131, *133*, *134*
Garnet, R., 30, *35*
Gentner, D., 15, *35*
Gilder, G., 60, 61, 63, *65*, 74, *83*

Gillan, J., 32, *35*
Goldberg, V., 42, 45, 54, *57*
Goldstein, M., 221, 227, 228n6, 230, *235*
Gonzalez, E., 196, *210*
Good, H. G., 24n6, 33, *35*
Gooding, R. Z., 221, 227, 228n6, 230, *235*
Gordon, K., 138n2, *156*
Gore, A., 192, 196, 197, 200, 207, *210*
Governor's Telecommunications Policy Coordination Task Force, 218, *235*
Groups petition FCC, 103, *107*
GTE Northwest Inc. v. Public Utility Commission of Oregon, 55, *57*

H

Hadden, S., 32, *35*
Hansen, W. L., 33, *35*
Haring, J., 47, *57*
Harrell, B., 45, *57*, 114, 118, *134*
Harris, D., 45, *57*, 114, 115, 116, 117, 118, 119, 120, 121, 122, 123, 124, 125, 126, 127, 128, 129, 130n4, 130n5, 131, *133*, *134*
Haviland, S., 60, 61, *65*
Hedemark, F., 45, *57*, 114, 115, 116, 117, 118, 119, 120, 121, 122, 123, 124, 125, 126, 127, 128, 129, 118, 130n4, 130n5, 131, *133*, *134*
Hillesheim, J. W., 23n5, 24, 24n7, *35*
Hong, P., 99, *107*
Horrigan, J., 191, *210*
Horrigan, J. B., 77, *83*
Horwitz, D., 239, 241, *248*
House Report on the Cable Communications Policy Act of 1984, 60, 61, *65*
Hudson, H., 32, *35*
Hudson, H. E., 32, *37*, 228, *235*
Hunter, H., 238, *248*

I

Illinois Commerce Commission (ICC), 245, *248*
Imafuku, H., 117, 129, *133*, *134*
In the Matter of The Bell Atlantic Telephone Companies, 138, *156*
Information Infrastructure Task Force (IITF), 32, *35*, 189, 192, *210*
Inman, S., 45, *57*, 114, 118, 120, 121, 122, 131, *133*, *134*
International Telecommunication Union (ITU), 190, *210*
Investigation into Nontraffic Sensitive Cost Recovery, 138, *156*
Iowa State Utilities Board, 180, *187*
Irving, L., 192, 196, 197, *210*
Itkin, L., 225n3, *235*

J

Jacobson, R., 31n10, *36*
Jaffe, A. J., 28, *36*
Jamison, M., 45, *57*, 114, 115, 116, 117, 118, 119, 120, 121, 122, 123, 124, 125, 126, 127, 128, 129, 130n4, 130n5, 131, *133*, *134*, 189, *212*
Jeziorski, M., 15, *35*
Johnson, F. G., 151, *156*
Johnson, M., 16, 29, 29n9, *36*
Johnson, M. A., 101, *107*
Jordan, K. F., 26, *35*

K

Kahn, A., 135, 136, 137, *156*
Karabel, J., 27, *36*
Kasserman, D. L., 135, 136, 137, 139, 152, *156*
Katz, J., 193, *210*
Katz, M. S., 25n8, 26, 33, *36*
Kedzie, C., 62, *65*
Kelley, K., 101, *107*
Kelsey, F., 189, *209*
Keltner, B., 62, *65*
Kirchhoff, H., 218, *235*
Kotin, L., 22, 23, 23n4, 23n5, 26, *36*

L

Lakoff, G., 29, 29n9, *36*
Lande, J., 189, *210*
Lavin, D. E., 27, 28, *36*
Law, S. A., 62, *65*, 144n7, 145n9, *156*
Lawton, R. W., 215, *234*
League of California Cities, 62, *65*
Leighton, W., 217, *235*
Little, L., 45, *57*, 114, 117, 119, 120, 121, 122, 123, 124, 125, 126, 127, 128, 129, 130n4, 131, *133*, *134*
Littlechild, S. C., 31n11, *36*
Lock, B., 117, 129, *133*, *134*
Louisiana Public Service Commission v. FCC, 180, *187*

M

MacMeal, H., 150, *156*
Makeeff, S., 45, *57*, 114, 115, 116, 117, 118, 119, 120, 121, 122, 123, 124, 125, 126,

127, 128, 129, 130n4, 130n5, 131, *133*, *134*, 189, *212*
Mandell, L., 141, *157*
Martin, P., 115, 116, 117, 119, 123, 124, 125, 126, 127, 128, 129, 130n4, 130n5, *133*, *134*
Marvin, C., 240, *249*
Massachusetts Department of Public Utilities, 152, *157*
Mayo, J. W., 135, 136, 137, 139, 152, *156*
McCarthy-Ward, P., 129, 130n4, 130n5, *133*
McConnaughey, J., 189, 192, 193, 196, *209*, *210*
McLaughlin-Krile, E., 225n3, *235*
McMillin, R., 119, 127, 128, *134*
McNagny, S. E., 102, *108*
Memorial Hospital v. Maricopa County, 74n13, *84*
Merrill, G. D., 23n5, 24, 24n7, *35*
Meyer v. Nebraska, 74n13, *83*
Miller, J. L., 33, *36*
Mitchell, B. M., 31n11, *36*, 62, *65*, 144n7, 145n9, *156*, 227, *234*
Monfils, J., 45, *57*, 115, 116, 117, 119, 120, 121, 122, 123, 124, 125, 126, 127, 128, 129, 130n4, 130n5, 131, *133*, *134*
Monroe, T., 45, *57*, 115, 116, 117, 118, 119, 120, 121, 122, 123, 124, 125, 126, 127, 128, 129, 130n4, 130n5, 131, *133*, *134*
Moore, G., 103, *108*
Morris, B., 142, *157*
Motorola goes for the hard cell, 145, *157*
Moyer, A. J., 151, *157*
MTS and WATS Market Structure, 180, *187*
Mueller, M., 30, *36*, 45, *57*, 86, *98*, 112, *133*, 167, 168, *177*, 191, *210*, 218, 230, *235*, 237, 241, 247, *249*
Mueller, M. L., 77, 78, *83*
Munday, L. A., 27, *36*
Murray, A., 59, 61, *65*

N

National Cable Television Association, 61, *65*
National Center for Education Statistics (NCES), 193, 194, 195, *210*
National Commission on Libraries and Information Science (NCLIS), 193, 194, *210*, *211*
National Communications Infrastructure (Part 3), 102, *108*
National Education Association and Princeton Survey Research Associates, 193, *211*
National Rural Telecom Association v. FCC, 185, *187*
National Telecommunications and Information Administration (NTIA), 32, *36*, 81, *83*, 191, 198-200, 207, *211*, 240, *249*
New study says target..., 102, *108*
Nila, C., 192, 193, 196, *210*
Nishioka, Y., 117, 129, *133*, *134*
Noam, E., 48, *57*
Noam, E. M., 215, 216, *235*
Noll, A. M., 229, *235*
Nowak, J. E., 71, 72, 73n8, *83*
NTIA study..., 101, *108*
Nuessel, F., 15, *36*
NYNEX accepts..., 101, *108*

O

O'Brien, M., 45, *57*, 114, 118, *134*
O'Connor, B., 32, *36*
Oettinger, A., 247, *249*
Office of Technology Assessment, 32, *36*, 194, *211*
Olivas, M., 33, *36*
(Online) NetDay 2000 Home Page, 197, *211*
Open video rules..., 103, *108*
Organization for Economic Co-Operation and Development, 145, *157*

P

Panis, C., 62, *65*
Panzar, J., 43, *57*, 168, *177*
Parker, E., 241, *249*
Parker, E. B., 32, *37*, 228, *235*
Parker, R. M., 102, *108*
Pearce, D. W., 63, *65*
Pearl, D., 229, *235*
Phillips, C., 247, *249*
Pingle, M., 242, *249*
Pitts, T., 115, 116, 117, 118, 119, 123, 124, 125, 126, 127, 128, 129, 130n4, 130n5, *133*, *134*
Pliskin, J., 62, *65*
Polanco, R., 103, *108*
Porter, K. H., 21, *37*
Povich, L., 191, *211*
President's Commission on Higher Education, 27, *37*

Q

Quality Education Data, Inc., 193, *211*

R

Ralston, L., 45, *57*, 120, 121, 122, 131, *133*
Rask, K. J., 102, *108*

Recommended Decision, 169, *177*
Redford, E., *79*, *83*
Response of Southern New England Telephone, 139n3, *157*
Rever, P. R., 27, *36*
Rhodes, L., 77, *83*, 191, *210*
Richards, L. E., 69, *83*
Richey, H. G., 26, *35*
Rizzo, C., 129, *134*
Rockman, S., 101, *108*
Roscoe, A. D., 32, *37*
Rose, N., 140, *157*
Rosenberg, E., 225, 226, 226n4, 227n5, *235*
Rosenkopt, L., 242, *248*
Rossman, J. E., 27, *37*
Rotunda, R. D., 71, 72, 73n8, *83*
Ruben, B., 101, *108*
Rubin, S. J., *77*, *83*
Rubin v. Coors Brewing Co., 73, *83*
Rural Utilities Service (RUS), 199, 208, *211*
Ryan, M., 99, *108*

S

Sawhney, H., 21n2, *37*
Schement, J., 191, *210*, *211*
Schement, J. R., 77, 78, *83*, 86, *98*
Schrag, P., 27, *37*
Schrage, M., 59, 61, *65*
Scotchmer, S., 139, *157*
Secretary of Commerce Daley, W., 207, *211*
Shankerman, M., 47, *57*
Shapiro v. Thompson, 73n8, *83*, 105
Sheppard, N., 101, *108*
Shew, W., 135, 136, 137, *156*
Shin, S., 238, *249*
Shriver, J., 100, *108*
Sichter, J., 45, *57*, 114, 117, 118, *133*, *134*
Silberstein, R. A., 27, 28, *36*
Sims, G., 45, *57*, 114, 115, 116, 117, 119, 120, 121, 122, 123, 124, 125, 126, 127, 128, 129, 130n4, 130n5, 131, *133*, *134*
Sloan, T., 192, 193, 196, *210*
Snider, J. H., 100, *108*
The Software Revolution, 143, *157*
Southwestern Bell Telephone, 191, *211*
Spielman, F., 101, *107*
Sprint Communications Company, L. P., 80, *83*
Sprout, A. L., 143, *157*
Stanford, J. D., 225, 226, 226n4, 227n5, *235*
Stigler, G., 140n4, *157*
Stoll, C., 230, *235*
Stone, A., 215, *235*
Strover, S., 228, *235*

T

Talley, W., 140, *157*
Tardiff, T. J., 153, *157*
Taylor, L. D., 31n11, *37*, 131, *133*
Taylor, R., 101, *108*
Taylor, W. R., 138n2, *156*
Teachers Network, 21st Century, 197, *211*
Tech Corps, "A One Page Perspective," 197, *211*
Telecommunications Act of 1996, 103, 105, *108*, 167, *177*, 179, *187*
Teske, P., 215, 216, *235*
Thompson, R. L., 20, *37*
Thomsen, K. A., 141, 142, *157*
Thönes, R., 45, *57*, 114, 118, *134*
TIIAP, Fact School, 197, *211*
Turner Broadcasting System v. FCC, 103, *108*

U

United States v. Winstar, 53, *57*
Universal Service Rules, 169, *177*
Urban, W., 23, 23n3, *37*
U.S. Bureau of the Census, 75n14, 76, 77, 78, *83*, 190, 192, *211*
U.S. Department of Agriculture (USDA), 190, *212*
U.S. Department of Education, 196, 199, *212*
U.S. Office of Education, 27, *37*
U.S. v. Western Electric, 245, *249*

V

Vasington, P., 117, 129, 130n4, 130n5, *133*, *134*
Vietor, R., 140, *157*
Vogelsang, I., 31n11, *36*, 227, *235*
Von Auw, A., 237, *249*

W

Wagoner, J., 23, 23n3, *37*
Warwick, M., 99, *108*
Weikle, J., 101, *108*
Weiman, D., 151, *156*
Weinhaus, C., 45, *57*, 114, 115, 116, 117, 118, 119, 120, 121, 122, 123, 124, 125, 126, 127, 128, 129, 130n4, 130n5, 131, *133*, *134*, 189, *212*, 247, *249*
Weisman, D., 47, 54, 55, *57*, *58*
West Lynn Creamery, Inc. v. Healy, 73, *83*
White House, 197, *212*
Wiggin, S., 42, *58*
Wildman, S., 43, 47, 50n2, *57*, 168, *177*

Wilhelm, T., 191, *212*
Williams, F., 32, *37*, 228, *235*
Williams, M. V., 102, *108*
Williamson, C., 21, *37*
Williamson, O., 42, *58*
Willingham, W. W., 27, *37*
Wisconsin Telephone News, 147, *157*
Wisconsin v. Yoder, 25, *37*
Witmer, D. R., 33, *35*
Wizdom Systems, Inc., 97, *98*

Y

Ying, J., 238, *249*
Young, P. R., 103, 108

Subject Index

A

Access costs debate, 135–155
  in an unregulated market, 149–155
  annual fees, 140–141
  competition as benchmark, 137–139
  increasing returns, 142–143
  pricing, 143–146
  segmented pricing, 139, 141–142, 144–145
  trends under competition, 146–149
  value-of-service pricing, 140, 146, 149–155
Agenda for Action, 189
America Speaks Out, 194–195
Ameritech, 10, 132, 159–164
  Customers First Plan, 164, 245–247
Anchorage Telephone Utility, 132
Annenberg Washington Program, 195–196
ASCII text, 87
AT&T, 132, 232
  divestiture of, 243–244
  enters credit card market, 141
  implements universal service in 1907, 237
  Presidential Conference (1996), 149, 152
  pricing, 10, 146–147, 149–154

B

Bell Atlantic, 132, 218
Bell Operating Company, 221–222
Bell telephone system, 30, 148t9.2, 149
  breakup, 4, 190
Bellcore, 132
BellSouth, 132
Benchmark Subsidy Method, 122–123, 123f8.10, 124f8.11, 125f8.12
Benchmarks, rate, 128f8.15, 128–129, 176–177, 184–185
Bilateral rules in regulations, 40, 41–42, 43–45, 44f3.1, 47, 48–52, 51t3.3, 54t3.4, 54–56
Broadband infrastructure, 128f8.15, 128–130, 129f8.16
BT, 132
Bulletin Board Services (BBS), 87

C

Cable television, 80, 101, 102–103, 191
Cable television companies (CATV), 242
California Higher Education System, 28, 34
Carrier of last resort (COLR), 46t3.1, 48–49, 49n1, 215, 227
Carriers, cellular, 242
Cell-phones, 145
Central processing unit (CPU), 93
Child labor, 25–26
Citizens for a Sound Economy, 217
City University of New York, 28, 34
Commerce clause, 54t3.4, 55
Commission on Higher Education, 27
Common carrier obligations, 46t3.1, 48–49, 49n1, 170–171, 174, 215, 224
Communications Act of 1934, 60, 62, 70, 80, 167, 179, 200, 240–241
Communications policy, 126, 127f8.14
Community computing, 85–98
  in an inner city, 94–98
  components of information access in, 91–94, 92f6.1
  demographics of, 89
  empowerment with information in, 91–94
  South Chicago Study, 89–94, 97–98
  today, 86–91

Competition
FCC role in, 177, 237
local, 237–248
local exchanges (LECs) as, 135, 237–38, 242–244
necessary to telephone network, 137–141
pricing under, 146–149, 161–164, 238–241
subsidies and, 102, 218, 222
in telecommunications markets, 11–12, 138–140, 216–219
universal service policy and, 4–5, 43–54, 47–48, 137–141, 146–153, 161–164
Competitive access providers (CAPs), 102, 237
Competitive equilibrium, 47
Competitive neutrality, 222–225, 231
Conference Report for the Telecommunications Act of 1996, 135
Connecting the Nation: Classrooms, Libraries, and Health Care Organizations in the Information Age, 196
Constitutional Limits on Government Action, 51t3.3
Contract clause, 51t3.3, 52, 52n6, 53, 54t3.4, 56
Corporation for Public Broadcasting, 132
Credit card companies, 140–141
Customers First Plan, 164, 245–247
Cybercafes, 88

D

Deaveraging telephone rates, 119–122, 120f8.7, 121f8.8, 122f8.9, 218–219, 244
Demographics
community, 89
of The South Chicago Study, 89–94, 97–98
universal telephone service and, 75–78
Detroit public schools, 87–88
Disconnection, a component of telephone service, 77–78
Due process clause, 51t3.3, 52n4, 52–53, 54t3.4, 55, 57

E

E-mail, 80
Economic value of a universal service policy, 72–74
Education, universal, 6, 16, 33–34. *See also* Schools
expansion of, 19, 22–28
higher education, 27–28, 33, 230
role of state in, 24–25
society's support of, 229–230
Educational institutions. *See* Schools
End-use devices, 93
Ex post facto clause, 51t3.3, 52, 52n7, 57
Expansion
demographic, 17–18, 18f2.2
layered, 18–19, 19f2.3
in a telephone network, 29–32
territorial, 16–17, 17f2.1
Expansion process, 19–28
in a telegraph network, 6, 16, 19–20
in universal education, 6, 16, 19, 22–28
in universal suffrage, 6, 16, 19, 20–22
Extended area service (EAS), 228

F

Falling Through the Net, 190–191, 196
FCC Report and Order, of the Federal Communications Commission (FCC), 169–77, 201–202
Federal Advisory Committee, 177
Federal Communications Commission (FCC), 11, 103–106, 132
adoption of new universal service rules (1997), 59
FCC Report and Order, 169–177, 201
Federal Advisory Committee, 177
implementor of telecommunication acts, 167–177
role in local competition, 237
universal telecommunications services and, 179–187, 216, 220–221, 222
Federal Home Loan Bank Board, 53
Federal–State Joint Board on Universal Service, 169–170, 177, 180, 184–185, 201, 220
Fiber-optic networks, 161
Financial Institutions Reform, Recovery, and Enforcement Act of 1989, 53
Freenets, 87

G

G.I. Bill of Education, 27
Graphical user interfaces (GUIs), 87–88
GTE, 132

H

Health care providers
Connecting the Nation: Classrooms, Libraries, and Health Care Organi-

zations in the Information Age, 197
Internet access for, 5, 201–202, 205–207
telecommunications services for, 4, 103–104, 168f11.1, 169, 172, 175–76, 201–202, 220

I

IITF Telecommunications Policy Committee (TPC), 194
Illinois, cost-based rates in, 243–47
Illinois Commerce Commission (ICC), 132, 243–47
"In groups," 17n1, 31
Independent telephone companies, 10, 147–149, 148t9.2, 150, 221–222, 223t14.1
InfoCom Research, Inc., 132
Information access
components of, 91–94, 92f6.1, 99–107, 220
cost of, 105
empowerment with, 91–94
"Information have nots," 9, 100–101, 105n1, 190–191, 192–193, 209
"Information haves," 105n1
Information Infrastructure Assistance Program (TIIAP), 198, 208
Information Infrastructure Task Force (IITF), 189
The information "revolution," 85–86
The Information Studio, 130
Inner city
"information have-nots" in, 100–101, 190–191, 192–193
resources, 94–97
Internet access
component of universal service, 80
"digital divide" and, 171–172, 193
for health care providers, 5, 201–202, 205–207
Internet service providers (ISPs), 143–145, 144t9.1, 176
for libraries, 5, 62–63, 192–194, 203–204
for low-income users, 196
Point-to-Protocol/Serial Line Internet (PPP/SLIP), 87
for rural areas, 204, 205–207
for schools, 5, 62–63, 193–194, 203–204, 208
Interstate long distance, subsidies for, 118f8.5
Intra Metropolitan Statistical Area (intraMSA), 244
Intrastate eligible telecommunications carriers (ETC), 224–225
Intrastate high-cost fund (HCF), 244
Iowa Utilities Board, 132
ISDN connectivity, 87–88

L

Libraries
Connecting the Nation: Classrooms, Libraries, and Health Care Organizations in the Information Age, 197
Internet access for, 5, 62–63, 192–94, 203–204, 208
telecommunications services for, 4, 103–105, 169, 171, 174–175, 192–194, 201–202, 203–205, 220
Lifeline, 42, 46t3.1, 168, 173–174, 203, 225–227
Link-up America programs, 42, 168, 173–174, 199, 203
Local exchange carriers (LECs), 42, 102, 247
competitive markets and, 135, 237–38, 242–244
monopoly of, 146–147, 149, 215, 237–238
operating revenues of, 129
service rates of, 180, 183, 185, 231–32
Long run service incremental costs (LRSIC), 246
Loop costs, 113, 135, 137–138, 154
subsidies for, 117f8.4
Low-income Internet users, 196
Low-income telephone customers, 146–147, 159–160, 168f11.1, 169–174, 190–194, 196, 201–203, 220, 225–226, 243

M

Market-driven economy, 106
Market failure, 103
Massachusetts Department of Public Utilities, 132
MCI, 146
MCI/Hatfield subsidy study, 113, 115f8.2
Media, growing importance of, 81
Minorities
electronic redlining of, 105
lacking telephone service, 76, 101, 103, 191, 196
in schools, 100–101
Mobile communications market, 145–46
Mobility, a component of telephone service, 77

Monopoly
local exchanges (LECs) as, 146–47, 149, 215, 237–38
universal service policy under, 39, 42, 45–48, 46t3.1, 113, 161, 215, 219, 222, 233, 238
Morse telegraph network, 20, 30
Multimedia, 99–100

N

National Commission on Libraries and Information Science (NCLIS), 193–194
National Education Technology Funding Corporation, 104–105
National Exchange Carriers Association (NECA), 132, 217
National Information Infrastructure (NII), 100, 189–209
Advisory Council, 189
Agenda for Action, 189
health care providers and, 201–202
importance of universal service to, 70–75, 81–82, 191–194
inner-city customers and, 100–101, 190–91, 192–193
libraries and, 192–94, 201–202, 203–205
low-income customers and, 196, 201–203
rural customers and, 190–194, 198–199, 203–204
schools and, 192–94, 203–205
National Regulatory Research Institute, 218
National Science Foundation, 94
National Telecommunications and Information Administration
(NTIA), 189
NetDay, 197, 199
Netscape, 142–143
Netscape Navigator, 143
New York Public Service Commission, 132
New York School Conference, 159
Nortel, 132
NTT America, 132
NYNEX, 132

O

Ohio Public Utilities Commission, 132
OPASTCO subsidy study, 115f8.2
"Out" groups, 17n1, 31

P

Pacific Telesis, 132
Pay phones, 101–102
"Plain old telephone service" (POTS), 189, 207, 209, 219, 240
Point-to-Protocol/Serial Line Internet (PPP/SLIP), 87
Political value of a universal service policy, 71–72
Postal telephone and telegraph entities (PTTs), 39, 41, 42–43
Poverty, *See also* Inner city; Low-income customers
component of telephone service, 76–77, 101–102, 103
electronic redlining and, 103, 105
President's Report on Higher Education, 27
Price averaging in telephone service, 119, *see also* Deaveraging telephone costs
Pricing
cost-based, 136, 150–154, 183–185, 216–219
increasing returns, 142–143
in mobile communications market, 145–146
segmented, 139, 141–142, 144–145
strategies in, 139–145
under telephone competition, 146–149, 161–162, 163–164, 238–241
value-of-service, 140, 146, 149–55
Primary Toll Carrier, 244
Privacy, a component of telephone service, 77
Private Branch Exchanges (PBX), 93
Public-Switched Telephone Network (PSTN), 237

R

Redlining, electronic, 101, 103, 105
Regional Bell Operating Companies (RBOCs), 218
Regulatory policies, social goals and, 40–43
Residential telephone customers, 120, 121f8.8, 160–61, 218–19, 222
FCC-defined services for, 202
Rural Electrification Administration and Rural Telephone
Bank, 217
Rural telephone customers, 102, 147, 162, 169, 171, 176, 198–199, 201–203, 220–221, 228, 243
broadband infrastructure for, 128f8.15, 128–129
impact of deaveraging on, 119–22, 120f8.7, 121f8.8, 122f8.9
"information have nots," 100–101, 190–194

Internet access for, 204, 205–207
Rural Utilities Service (RUS), 198–199

S

Santa Monica Public Access Network, 87
Santa Monica Public Education Network (PEN), 87
SBS Communications Inc., 132
Schools
broadband service for, 129–130
Connecting the Nation: Classrooms, Libraries, and Health Care Organizations in the Information Age, 197–198
information access for, 104–105
Internet access for, 5, 62–63, 193–194, 203–204, 208
minorities in, 100–101
telecommunications services for, 4, 103–105, 168f11.1, 169–175, 192–194, 201–205, 220
Social goals
economic activity and, 44f3.1
regulatory policies and, 40–43
Social value of a universal service policy, 40–43, 74–75, 78–79
The South Chicago Study
demographics of, 89–94, 97–98
Southern Bell, 149
Southwestern Bell, 115f8.2
Special-needs telephone customers, 159–160
Sprint, 132, 146
Sprint Cellular, 132
Sprint Local Telecom Division, 132
Sprint/Sievers study, 115f8.2
States
affordable rates in, 220–221
broadband infrastructure in, 128f8.15, 128–130, 129f8.16
competitive neutrality in, 222–225, 231
cost-based rates in Illinois, 243–247
deaveraging rates in, 218–219
extended area service (EAS) in, 228
independent telephone companies in, 221–222, 223t14.1
Lifeline programs in, 203, 225–227
local exchange carriers (LECs) in, 215, 231–232
monopoly franchises in, 215, 219, 222, 233
role in telecommunications services, 215–234, *see also* Federal–State Joint Board on Universal Service
subsidies in, 217–219, 224–226
universal service funds in, 225–227
virtual vouchers for service in, 230–231
Subsidies
Benchmark Subsidy Method, 122–123, 123f8.10
benchmarks for, 176–177, 184–185
competition and, 102, 218, 222
differing amounts of, 116f8.3
for monthly loop costs, 117f8.4
new structures for, 117–127, 119f8.6, 124f8.11, 125f8.12, 126f8.13
new thinking needed, 219
for telecommunications services, 47–48, 101, 112–117, 114f8.1, 116f8.3, 160–163, 204–207, 217–19, 224–226
USTA/Monson–Rohlf subsidy study, 113, 115f8.2
virtual vouchers as, 230–231
Suffrage, universal, 6, 16, 19, 20–22
Supremacy clause, 51t3.3, 52, 52n5, 54t3.4, 55–56, 57
Survey of Rural Information Infrastructure Technologies, 198

T

Takings clause, 51t3.3, 52n3, 52–53, 54t3.4, 55, 57
TECH CORPS, 197
Technology Innovation Challenge grants, 199
Technology Literacy Challenge initiative, 199
Telecommunications Act of 1996, 4, 10–12, 39, 59, 103–106, 167, 215, 216–219
promotes competition, 138
provisions of, 200–201
signed into law, 200, 215
statutes of, 179–187
Telecommunications Industries Analysis Project Interstate study, 115f8.2
Telecommunications Industries Analysis Project (TIAP), 111–112, 114f8.1, 130, 132–133
Telecommunications markets, competition in, 11–12, 138–140, 216–219
Telecommunications policy, availability and affordability as, 201–202, 220–221, 240–241

Telecommunications relay service (TRS), 227
Telecommunications services
carriers, 48–49, 49n1, 181, 215, 227, 242
debates over service, 111–112
for health care providers, 4, 103–104, 168f11.1, 169, 172, 175–176, 201–202, 220
increasing competition in, *see* Unilateral rules, Bilateral rules
for libraries, 4, 103–105, 169, 171, 174–75, 192–94, 201–202, 203–205, 220
overview of current subsidies, 112–117
for schools, 4, 103–105, 168f11.1, 169–75, 192–94, 201–205, 220
states' role in, 215–234, *see also* Federal–State Joint Board on Universal Service; States
statutory provisions for universal, 179–187
subsidies for, 47–48, 101, 112–117, 114f8.1, 116f8.3, 160–163, 204–207, 217–219, 224–226
Telegraph network, 6, 16, 19–20
Telephone network
affordable rates, 201–202, 220–221, 240–241
Bell system, 4, 30, 132, 148t9.2, 149, 190, 218, 221–222
central office equipment (COE), 244
competition in, 4–5, *see also* Competition
competitive neutrality in, 222–225, 231
cost-based rates, 136, 150–54, 183–185, 216–219, 243–247
deaveraging costs in, 119–122, 120f8.7, 121f8.8, 122f8.9, 218–219, 244
demographics of, 75–78
expansion of, 29–32, 62–63
FCC-defined services for households, 202
local exchange carriers, *see* Local exchange carriers (LECs)
loop costs, 113, 117f8.4, 135, 137–138, 154
low-income telephone customers, 44t3.1, 76–77, 146–147, 159–160, 168f11.1, 169–174, 190–194, 196, 201–203, 220, 225–226, 243
minorities underserved in, 76, 101, 103, 105, 191, 196, 201
national information infrastructure component, 100
penetration rates, 190–191, 192–193
pricing under competition, 146–149, 161–164, 238–241
privacy a component of, 77
Public-Switched Telephone Network (PSTN), 219
residential customers, 120, 121f8.8, 160–161, 202, 218–219, 222
rural customers, 102, 119–122, 120f8.7, 121f8.8, 122f8.9, 128–129, 147, 162, 169, 171, 176, 190–194, 198–199, 201–204, 220–221, 228, 243
rural wire centers, 219
special-needs telephone customers, 159–160
universal availability of, 8, 101, 112–113
urban customers, 120f8.7, 121f8.8, 122f8.9, 128f8.15, 128–129, 160–162, 168, 175
urban wire centers, 219

U

Unilateral rules in regulations, 40, 41–42, 43–45, 44f3.1, 47–48, 50t3.2, 51t3.3, 53–54, 54t3.4
Universal service, *see also* Information access; Internet access
Ameritech's response to, 159–164
conceptualizing, 7–12
definition of, 168, 238–241
demographics and, 75–78, 89–94, 97–98
expansion metaphors for, 19–32
important to National Information Infrastructure (NII), 70–75, 81–82
to include broadband, 128–130
information policy and, 80–82
National Information Infrastructure (NII) and, 189–209
telecommunications implications for, 32–34
Universal Service Fund (USF), 217
Universal service overview, 111–133
current subsidies, 112–117
implications of including broadband, 128–130
need for new structures, 117–127
Telecommunications Industries Analysis Project, 111–112, 114f8.1, 130, 132–133
Universal service policy
competition and, 43–54, 47–48
competitive equilibrium, 47
components of a new, 78–81
implementation of, 61–64, 103–104

in the information age, 191–94
justifications for, 71–75
under monopolies, 39, 42, 45–48, 46t3.1, 113, 161, 215, 219, 222, 233, 238
need to include cost-based rates, 243–247
need to reflect individual jurisdictions, 241–243
political value of, 71–72
rationale for, 59–61
review of Federal role in, 167–177, *see also* entries under Federal...
social value of, 40–43, 74–75, 78–79
Universal Telephone Service Protection Law of 1985 (UTSPL), 244–245
UNIX, 89
Urban telephone customers, 120f8.7, 121f8.8, 122f8.9, 128f8.15, 128–129, 160–162, 168, 175
Urban wire centers, 219
US WEST, 132, 232
USTA Member Study, 115f8.2
USTA/Monson–Rohlf subsidy study, 113, 115f8.2

V

Video
dial tone networks, 103
information, 102–103
Virtual community, 88
Virtual vouchers, 230–231

W

Washington State Utility and Transportation Commission (WUTC), 218
Washington Utilities and Transportation Commission, 132
Western Union, 146
Wizdom Systems, Inc., 97
Working Group on Universal Service, NTIA, 194
World Wide Web (www), 87